"A fascinating book . . . There is an energy in the book and in Ghonim's words that makes one feel it is much too soon to assume the revolution is over, or to underestimate what the rebels achieved."

—*Philadelphia Inquirer*

"A welcome and clear-eyed addition to a growing list of volumes that have aimed (but often failed) to meaningfully analyze social media's impact. It's a book about social media for people who don't think they care about social media. It will also serve as a touchstone for future testimonials about a strengthening borderless digital movement that is set to continually disrupt powerful institutions, be they corporate enterprises or political regimes . . . [Ghonim's] individual story resonates on two levels: it epitomizes the coming-of-age of a young Middle Eastern generation that has grown up in the digital era, as well as the transformation of an apolitical man from comfortable executive to prominent activist."

—*New York Times Book Review*

"Deserve[s] to become part of the canon of classic prison literature."

— *Washington Post*

"An articulate account of the author's middle-class upbringing under a draconian regime, and a gripping chronicle of how a fear-frozen society finally topples its oppressors with the help of social media . . . *Revolution 2.0* reminds us that anonymity is crucial when one lives in a nation where state security has a dossier on every citizen and the license to detain and mistreat on any grounds."

—*San Francisco Chronicle*

"Ghonim tapped into a demographic that proved crucial to the Egyptian uprising: upwardly mobile college-educated youth frustrated by Egypt's stagnation but wary of politics and activism . . . *Revolution 2.0* narrates a fascinating but narrow slice of the process through which apathetic and passive people are politicized."

—***Daily Beast***

"*Revolution 2.0* . . . is likely to be required reading for web geeks, media experts, political scientists, advertising executives, activists, anarchists, confidence men, secret policemen, dictators and corporate strategists."

—***Telegraph*** (UK)

"A remarkable personal testament that will be cited by future historians of both Facebook and the Arab Spring."

—***Kirkus***

"There's no doubting that his tell-it-like-it-is memoir will be studied by historians for generations to come."

—***Newsday***

Revolution 2.0

THE POWER OF THE PEOPLE
IS GREATER THAN
THE PEOPLE IN POWER

A MEMOIR

WAEL GHONIM

MARINER BOOKS
HOUGHTON MIFFLIN HARCOURT
BOSTON • NEW YORK

First Mariner Books edition 2013

www.hmhbooks.com

Library of Congress Cataloging-in-Publication Data
Ghonim, Wael, date.
Revolution 2.0: the power of the people is greater than the people in power : a memoir / Wael Ghonim.
p. cm.
Includes index.
ISBN 978-0-547-77398-8 ISBN 978-0-547-86709-0 (pbk).
1. Ghonim, Wael, date. 2. Political activists — Egypt.
3. Internet — Political aspects — Egypt. 4. Egypt — History — 1981–
I. Title. II. Title: Memoir and call to action.
DT107.828.G445A3 2012
962.05'5092 — dc23
[B] 2011042557

Book design by Brian Moore

Printed in the United States of America
DOC 10 9 8 7 6 5 4 3 2 1

To all the patriotic Egyptians, Tunisians, and the rest of the Arabs who took over the streets and made history that inspired the world

To every brave man and woman who made the ultimate sacrifice

To future generations, in hopes that they will live in a free and democratic world

Contents

Revolution 2.0

Prologue

THE WORLD AROUND ME was reduced to pitch black. I could sense the deliberate use of side streets by the driver as the car traveled through Cairo at midnight. We twisted and turned many times, a technique my captors often used to disorient their victims.

On my right and left were two guards from State Security. They kept a tight grip on my handcuffed arms. I remained completely silent so as not to provoke them. They had forced my shirt up to cover my head so I could not see, and my belt was tied firmly over the shirt, around my head. One of them had pushed my head down to hide me from passing pedestrians. Everything I had been carrying had been confiscated.

Those brief moments before the car reached its destination were all too familiar. I had published the accounts of many captives of State Security. Now it was my turn. I wondered what could happen to me next, but I knew the answer: anything.

"Get out, you son of a b———," said a loud and angry voice when we arrived. I was being pushed out of the car. My reception inside the building was harsh and mocking. I was slapped, kicked, and cursed, all accompanied by derisive laughter. It seemed as if these men enjoyed their work, or at least they did it purposefully. The laughter was part of their strategy to instill fear prior to interrogating newcomers. The most difficult thing about the slaps and kicks was their element of surprise. I had no means of anticipating any strike because I was blindfolded. When would I be hit next? From which side, on which part of me? I had no clue.

I wondered what they knew. What had I done that had given me away? *Kick. Curse.* My fear grew. I knew that this was what they wanted—to break me down before the interrogation. I decided to hasten things along by pretending to tremble. Yet real fear was starting to take over.

In the midst of the beating I prayed to God that he would somehow inspire my friend Najeeb, in Dubai, to change the password to the Facebook page's e-mail account. I prayed for Najeeb to do it before the interrogation got serious. They must not know what I had done.

I wanted to see my children again.

1

A Regime of Fear

MY 2011 ARREST WAS not the first time I had encountered Egyptian State Security. One winter afternoon in 2007, I received a call from a man who presented himself as Captain Raafat al-Gohary, from the bureau in Giza, Egypt's third largest city, which is part of greater Cairo. Needless to say, Rafaat al-Gohary was not his real name. State Security officers feared the potential wrath of citizens they interrogated and tortured, so they used pseudonyms. I greeted him calmly, attempting to hide the anxiety caused by the surprise. He said I needed to meet him for an important matter and I was to head to State Security in Dokki, a neighborhood in Giza, at eleven o'clock that night. My anxiety increased. I asked what was the matter. His response: "There's nothing to worry about. We'll just have a chat over coffee, that's all." This failed to comfort me. I asked if we could reschedule, saying that I was busy with work. He refused. I wanted to play for time to try to figure out why I was being summoned, but he insisted we meet at eleven. *What is the worst that can happen?* I wondered. My days of activism were long over. I had never before been summoned.

Immediately after hanging up, I contacted a close friend, and we agreed that I was to call him right after the meeting ended. If he never got the call, he was to find out exactly what had happened to me, since in the past, people in a situation like mine had suddenly disappeared for days or even months after their "visit." I decided not to tell my wife or my family anything, as I didn't want them to panic.

I arrived at the main gate at 11 P.M. sharp. The neighborhood was quite familiar to me; my high school was literally right around the corner. At reception, after confirming that I was to meet Captain Rafaat al-Gohary, I was told to sit down and wait. Around me were at least six others. Although I didn't speak to them, it was clear that we all shared one emotion: apprehension.

Egyptian State Security reached deep into society, involving itself in every detail of life. It thrived on the emergency law, enacted in 1958 but not enforced until after the Six-Day War in 1967, and still in effect in mid-2011. That law gives executive authorities the right to arrest, interrogate, and imprison any Egyptian for up to six months without a warrant or any legal grounds or even the right to an attorney. It also empowers the authorities to ban all types of protests as well as gatherings of any group of people without a security clearance.

The dossiers of State Security were objects of fear and ridicule. Any activist of any sort, or even anyone with considerable financial or intellectual influence, had an exhaustive dossier in his or her name at State Security, containing every detail the authorities had collected that could possibly be useful in blackmailing him or her into obedience when needed. Privacy was almost meaningless to this quintessentially Machiavellian organization. Thus, phone tapping, for instance, was a very common practice of State Security officers. Word spread that tapes documenting the infidelities of famous businessmen and public figures were stored in a room at headquarters. Ironically, officers used to advise each other not to spy on their own wives' phones, to avoid family conflicts.

Not only did the state monitor and terrorize political opposition groups and religious activists, but its oppressive reach extended to anyone engaged in public service, including charities whose field operations were limited to empowering the poor and unfortunate. With over 40 percent of Egyptians living below the poverty line, the authorities were consistently trying to curb anyone who might mobilize the masses for a future political cause.

State Security approval was obviously a prerequisite for any senior appointment in the government. Even university teaching assistants, who are supposedly selected from among the top students of the year's graduating class, could not be hired by the university without a security clearance proving that they were innocent of any dissident activism, political or religious.

The Egyptian regime lived in fear of opposition. It sought to project a façade of democracy, giving the impression that Egypt was advancing toward political rights and civil liberties while it vanquished any dissidents who threatened to mobilize enough support to force real change.

The Ministry of Interior was one key force of coercion. Another was the state media: terrestrial and satellite television as well as newspapers and magazines, the most famous of which were *Al-Ahram, Al-Akhbar,* and *Al-Gomhouriya.* The regime sought to plant fear in the hearts of Egyptians from an early age. Fear was embodied in local proverbs, such as "Walk quietly by the wall (where you cannot be noticed)," "Mind your own business and focus on your livelihood," and "Whosoever is afraid stays unharmed." The regime's uncompromising control also covered workers' unions and the nation's legislative bodies.

This all amounted to what I came to call "weapons of mass oppression." No matter how far down we spiraled, no matter how much corruption spread, only a few people dared to swim against the current. Those who did ended up in a prison cell after an unfriendly encounter with State Security, or were subjected to character assassination in the media, or were targeted on fraudulent charges or long-ignored violations.

"Hello, Wael. Why are you giving us a hard time? Why the troublemaking?"

This, together with a faint smile, was how Captain Raafat greeted me. His air-conditioned office contained three other investigators. The room was modestly decorated with a number of books, many of which were very obviously about religion. State Security wanted everyone to believe that it had nothing against faith.

I looked at him and smiled as I responded calmly, "I don't make trouble at all. It is you guys who give *me* trouble, and I have no idea why. I'm glad you called me in, so I can figure out what the problem is. Every time I travel back to Egypt my name appears on the arrivals watch list and the airport officers transfer my passport to State Security, who pulls me aside for an inspection, including a full search of my bags."

This problem dated back to December 2001, when I returned from the United States, three months after 9/11. As I was collecting my luggage, I heard my name over the loudspeakers. I was urgently asked to return to passport control. There was also someone calling my name in person, so I showed myself to him. He took my passport and asked me to wait in front of a lounge by State Security's airport office. After a very nerve-racking forty minutes, a detective emerged with my passport and asked me to bring my luggage in for inspection. That day I thanked God that everything turned out well. It appeared to be nothing more than a typical post-9/11 glitch. Yet every time I entered Egypt between that day and the time the revolution began, I was pulled aside. Until this day, I had never found out the reason for that.

Captain Raafat was deliberately friendly, as if we really were just having a chat. However, he was armed with pen and paper, and he carefully documented the conversation. He took time to finish recording my responses before he resumed his questions. Almost everyone from the upper or middle class who was called in for interrogation by State Security was met with this same friendly, off-the-record manner. (Poorer people were treated far more harshly.) It was transparently illegitimate.

The captain asked for my personal information: name, age, address, marital status. I answered all his questions. He asked about my wife's full name.

"Oh, she is not Egyptian. Where is she from?"

"America," I responded.

He wrote her full name in Arabic as I pronounced it again and asked me to verify the spelling.

"So you married an American for the citizenship, right?"

He was surprised to discover that despite my marriage in 2001, I had never applied for a green card or U.S. citizenship. "I'm a proud Egyptian and I find no reason why I should apply for any other citizenship," I explained.

Very cynically, he replied, "And what is it exactly that you like about Egypt?"

"I'm never able to verbally express my reasons for loving Egypt, yet love for it runs in my blood," I replied honestly. "Even my wife asks why I love my country despite all its shortcomings. I always answer that I don't know why. You know, Captain, when I lived in Saudi Arabia, during the first thirteen years of my life, I literally used to count the days left, on a paper on my desk, before I could return home to Egypt to spend the annual vacation. And when only a few days remained, I was too excited to fall asleep at night." I returned his cynical smile and joked, "I love it here because life lacks routine. You wake up in the morning and have no idea what the day will be like. One morning you could receive a phone call like the one I received today, asking you to report to State Security."

He smiled while saying, "You are certainly a troublemaker."

I saw a copy of the Holy Qur'an lying on the captain's desk. I assumed it was there to assure anyone who sat opposite him that the captain regularly read scripture and had nothing against faith. The ruling regime was extremely apprehensive about organized religious forces in Egypt, particularly ones that concerned themselves with public affairs. Their fears were intensified when thousands of Egyptians traveled to Afghanistan to fight the Soviet invaders. Many of those fighters, or self-proclaimed mujahideen, returned with ideologies that rejected the Arab regimes, denouncing them as heretical and treacherous tools of the West. The new ideology, and the new militants, posed a threat to the Egyptian authorities. Although the emergency law had been suspended by President Anwar al-Sadat in 1980, it was reinstated eighteen months later, following Sadat's 1981 assassination at the hands of radical Islamists. Sadat's assassins were apparently motivated by his crackdown on more than 1,500 political and religious activists, and also by the fact that he signed a peace treaty with Israel and emphasized it with a visit to Tel Aviv.

The influence of religious groups in Egypt increased as time went on, and their variety expanded. These groups were never homogeneous, nor did they all necessarily share the same philosophies or even objectives. They did share one thing, however: enmity toward the regime. In turn, Hosni Mubarak's government feared them. Mubarak knew these groups could influence the Egyptian masses more than anyone else, since Egyptians tend to be religious by nature; in a Gallup poll conducted in June 2011, 96 percent of the one thousand Egyptian respondents agreed that religion played "an important role in their daily life." Ordinary Egyptians take religious figures as role models, symbols of nobility and sincerity, values which were thoroughly lacking in many of the public representatives of the regime. Most of the time when the regime attacked a religious group, that group's popularity received a boost. The fact that economic conditions were stagnant or declining only magnified the effect.

State Security kept an eye on all religious speakers and scholars and even on university students who frequented mosques, not just those who were active in Islamic movements. They were careful to summon such people to their offices to ask them about their activities and even to intervene and attempt to redirect them. Occasionally, hundreds would be arrested and thrown into jail for years without explicit accusations. Behind bars, they were brutally treated and humiliated. Once released, they either became fanatics, motivated by their bad experience, or attempted to reintegrate into society and forget the past.

This, I realized, was the real reason for my interrogation. State Security wanted to know if I had any links to religious or political activism, especially now that I regularly traveled abroad and, as a result, was becoming more exposed to real democracy. It was time to create a dossier in my name that contained the details of my life for future reference.

The story of my faith dates back to high school days. I did not pray regularly before then, although I adhered to the general ethics of religion, thanks to my parents' encouragement and because I grew up in Saudi Arabia. That country is conservative by nature, especially in Abha, a small southern city where society and culture are assumed to be less advanced than in urban centers.

One of my closest cousins, Dalia, died in a car accident in 1997 at the age of twenty-five. Her death had an impact on me, and I was moved to explore my faith, as I didn't want to die unprepared. I listened to sermons, attended religious lessons, and read books. I felt that life was a brief test that ended at death. I started praying five times a day, on time, and often at the mosque.

At the university, I mixed with people from many religious groups and ideologies, including the Muslim Brotherhood, and I joined many of their activities at the school. But I always made my own sense out of things. A famous sheikh whom I met with several times once said to me, "Your problem, Wael, is that you only follow your own logic and you don't want to have a role model to follow." It was hard for me to accept conventional wisdom. It was my nature to discuss any matter thoroughly before I could accept a conclusion with both heart and mind. This attitude in an eighteen-year-old is not always endearing. It was not just my age, however. Thanks to frequent exposure to global media and modern communication tools, many young Egyptians were slowly becoming empowered to make their own educated choices.

"So your dad lived in Saudi Arabia. For how many years? What are his religious and political views?" asked Captain Rafaat, who had to gather as much information as he could, not only about me but also about my family members, as part of his job.

My father is a typical hardworking Egyptian who comes from the slowly eroding middle class. Born in the 1950s, his generation sang praises to Arab nationalism and the 1952 revolution, when Egypt's

King Farouk was overthrown by a military coup and Egypt was transformed from a monarchy into a republic. My grandfather, may he rest in peace, was a government employee at the Egyptian Railways. He had seven sons, whom he struggled to raise and educate. My father, the oldest, graduated from medical school and immediately went to work for a public hospital.

After my father married my mother, in 1979, and I came along, in 1980, his salary could hardly cover our basic needs as a family, so he decided to leave for work in Saudi Arabia. It was a very tempting option for many Egyptians. The salary offered in Saudi Arabia was twenty times the amount he received at the public hospital in Egypt. Like millions of Egyptian expatriates, he hoped to save some money and then return home after a few years to start a private practice in Cairo. Egypt's talented citizens were becoming its main export, to the country's detriment.

Economic conditions at home were horrendous at the time. Every year tens of thousands of Egyptians applied to the green card lottery, hoping to emigrate to America. Others left for Gulf countries, Canada, or Europe, by any means possible, to look for job opportunities. The phenomenon kept increasing, and emigration became the common dream of scores of Egyptians. Those with fewer skills did not have as many options. Some were desperate enough to put their lives at risk by emigrating to Europe illegally, by boat, despite the risk of drowning. I still remember an Egyptian comedian's response to a question about the future of the nation: "Egyptians' future is in Canada."

After spending only a few years in Saudi Arabia, my father, like many Egyptians, fell into the trap of Islamic private investment companies, which proliferated in the early eighties. These companies offered a huge annual return on investment that reached 30 or sometimes even 40 percent, as opposed to banks, which offered 10 percent or less. My father deposited his life savings with four of these companies to diversify his portfolio. The companies were founded by religious Egyptians who offered their services as an alternative to banks;

various Islamic scholars deemed fixed interest rates to be usurious and consequently prohibited by Sharia law.

A few years after the enormous growth of these companies, the Egyptian regime decided to fight them. Among other things, it wanted to protect the interests of loyal businessmen and feared that these private asset management companies would control the economy and cripple the banks. All such companies were frozen by the state, and their founders were arrested for fraud and money laundering. Most of the money saved by my father after years of hard work in Saudi Arabia was lost, as was the money of many other middle-class Egyptians inside Egypt and abroad.

So my father decided to stay in Saudi Arabia for a much longer time than he had initially planned. Every time I asked him why we were not returning home, he would answer, "How can I provide for a family of five with a salary of a few hundred pounds that runs out by the fifth day of the month?" My father is typical of his generation. He is fun, everyone loves him, and back then he spoke about politics only through jokes that timidly criticized the ruling class. "Ignore, live, enjoy" was his philosophy. Whenever he could, he would ignore problems rather than face them. I don't blame him; the 1952 revolution had this effect on most of his generation.

My mother, on the other hand, pressured my father every year to return to Egypt, start his private practice, and attempt to readapt to life at home. We finally decided as a family that everyone but my father (I now had a brother and a sister) would return to Egypt and that he would follow us two or three years later, when he had saved enough to start a business at home. (Unfortunately, this never actually happened, and my father still lives in Saudi Arabia.)

Captain Rafaat was not very interested in my father once he found out that he was not involved with any political or religious groups, and he quickly moved on to ask, "So, when did you return to Egypt?"

It was in 1994. I enrolled in a private school in Zamalek, near our home in Mohandeseen. Both neighborhoods are known to be among the best areas in Cairo. I was in the ninth grade at the time. The decision to return to Egypt was one of the happiest moments of my life, but it was not easy living away from my father. I was never very capable of expressing emotions. I missed him immensely and always looked forward to his visits home. When he came home for forty-five days of vacation every year, I accompanied him everywhere he went. I laughed at his constant jokes and loved his modesty and his openness toward everyone he met. Tears always came to my eyes when he was leaving to go back to Saudi.

My mother did her best to make up for Dad's absence. She was fully devoted to raising her three children to become decent and responsible human beings, and I was impressed at how she selflessly agreed to be away from her husband in order to do so. Despite her incredibly strong character, she put her children first in every decision she made.

Fortunately, I quickly adapted at school. My best friend was a genius of a boy by the name of Moatasem. He always effortlessly came in at the top of our class. I tried competing with him during exams, but always in vain. Moatasem was extremely diligent. I scored 92.5 percent and ranked second after him in the ninth grade, which is a milestone year in our educational system, the final year before "secondary education." Moatasem decided to transfer to a public high school, where he would enroll in classes for advanced students. He convinced me to leave our private school and go with him to Orman High School. "It will be very competitive for us in the advanced classes, and the teachers in these classes are some of the best in Cairo," he said. These arguments were enough to convince me, but one more reason was to get to know the real Egypt and integrate with Egyptians from different backgrounds and social classes and not just those who could afford to go to private schools.

I missed the aptitude tests for the advanced classes because I was away on our annual visit to my dad in Saudi Arabia during the summer of 1995. Before I began traveling, an admissions employee at the school assured me that I would be able to take the aptitude test once

I returned. Unfortunately, however, he didn't keep his promise, so I found myself attending regular classes.

Orman High School gave me culture shock. It was worse than anything I had ever imagined or heard about public schools. Being an all-boys school, there was a constant surplus of testosterone in the air. Fighting in the school playground always ended with someone injured. There was a designated corner for smoking cigarettes, and sometimes hash. Skipping school was common, as long as you paid a toll—a bribe—to the student guarding the fence. The number of students in a single class was at least double what I had been used to, over seventy students in a space that had contained only thirty students at my previous school.

I quickly tried to reverse my decision by calling the principal at my previous school. He refused to take me back, in order to teach me a lesson: he had offered many enticements to keep me at the school when I announced my decision to transfer, including slicing my tuition fees in half. I was very stubborn and rejected all his offers, so I don't blame him for refusing me when I suddenly tried to crawl back. Unwittingly, however, I had made one of the most important decisions of my life.

It was no easy task to cope in the new environment. Blending in was more challenging to me than performing well in class, and I regained my balance only after I began to adapt. At the beginning of my Orman experience I hated it so much. Yet at the end I loved it just as much. That school exposed me to social classes I had never mixed with. I learned how to relate to all kinds of people. I later became extremely interested in psychology and sociology, not least because of these years.

In my first year, I received my worst grades ever. The threat of failure has always motivated me to fight back. I decided to focus all my time and effort during the next year—the eleventh grade—to excel, in order to join the advanced classes with my friend Moatasem in twelfth grade, the last year at high school. Mission accomplished: after a year of very hard work, I received a grade of 95 percent and was able once again to sit at a desk with Moatasem, as we used to do in the ninth grade.

Nevertheless, no amount of success could make me forget some of the things I saw during the first two years at Orman. The teachers tried to maintain order by means of violence and beatings. In return, the students enjoyed intimidating and harassing the teachers. There were daily battles in those classrooms of seventy, among whom were a fair number of troublemakers.

Like other government employees, public school teachers in Egypt receive a monthly salary of no more than a few hundred pounds, which does not cover their basic family needs. As a result, private lessons have become teachers' main source of income. Teachers can generate thousands of pounds by visiting students' homes and tutoring them in a far better environment than at school. A survey carried out by the Egyptian cabinet's Information Center in 2008 revealed that 60 percent of parents sought private lessons for their children. Many families were spending up to a third of their income on these lessons.

Like a cancer, the phenomenon of private lessons quickly spread everywhere in the country. Teachers began marketing their services on leaflets that can be found in every street of every city and town. They give themselves catchy titles like "the emperor of physics" or "the colonel of chemistry." The real shame is that most teachers, along with the government's textbooks, emphasize rote memorization rather than any genuine understanding. Students and parents have to find their own ways to learn how to solve problems. Many students rely on supplementary texts. Egyptians spend over one billion pounds ($200 million) every year on them. I resisted private lessons adamantly until my final and decisive year in high school, when math and chemistry were so challenging that I simply could not grasp them from the classroom instruction.

One of my elected courses was psychology. I chose to study it because, like many adolescents, I was interested in understanding human nature. I decided to take private lessons with a university instructor whom I will never forget: Mr. Ehab. We used to spend hours more than the scheduled time discussing many interesting topics. Mr. Ehab taught me how to deal with various people and situations and helped me realize that a large number of conflicts result from pure miscommunication, like what Aristotle said about the importance of defining terms to avoid unnecessary disagreement. It was quite a good experience for someone of my age.

The corrupt educational environment also encouraged cheating. Teachers who supervised without allowing cheating were described by students as "bothersome." Some mothers used to wish that the proctors of their children's exams would let them cheat. It is not surprising that cheating and fraud gradually became everyday activities in Egypt, making their way from education to business and commercial transactions, and ultimately to elections.

I graduated from high school with a total grade score of 97 percent. I was going to attend Cairo University to study engineering, but first I searched for a job. My primary reason was to pay my phone bill, which had soared for a reason my father might never have imagined: dial-up Internet access. I spent hours exploring the Internet, browsing websites and chatting anonymously with people I did not know from around the world, using mIRC (a famous chat client at the time) to make virtual friends. I remember when my dad stormed into my room during the summer after high school to express his anger at the size of the phone bill. He confiscated the computer and locked it up in a closet, explaining that I was irresponsible and that my relationship with the computer had to end. As soon as he left the house, I broke open the closet and reclaimed the computer. When he returned, I begged his forgiveness and declared that I would get a dedicated phone line and the bill would be my responsibility. Luckily, my father always tried to treat his children as responsible near-equals. He often told us to be careful what we wished for. This time, after hearing me out, he said, "As you wish." It was the beginning of my life online, and the beginning of my financial independence, as I started earning a steady income from working in a video gaming store and as a freelance website developer.

Working and spending long hours online was a real challenge to my studies. After passing the preparatory year at the engineering school, students were expected to choose a department to enroll in. The number of seats was limited in some departments, making them very competitive. I scored badly during my preparatory year in 1998. As a result, I initially enrolled in electrical engineering instead of my first choice, computer engineering. Nonetheless, I quickly determined that I really wanted to work with computers. A friend of mine had said that if I failed my first year in electrical engineering I could submit an appeal to the dean explaining that my life's dream was to study computer engineering, so I proceeded to Student Affairs, where I learned that my friend's information was accurate enough but success depended on the number of transfer requests submitted.

I took the risky decision to skip that year's exams and submit an appeal at the end of the year. As usual, my parents were surprised by my decision and tried all forms of dissuasion, but I insisted. After few months my wish came true: only one other student requested a transfer, and we were both admitted to computer engineering.

Life was different inside my new department. There were no more than forty students, and the professors and teaching assistants knew each one of us by name. I tried to compete with the top students, but I was always behind, thanks to the countless hours I spent online. I remember one teaching assistant, Ahmed, who paused during one of his lectures and singled me out. "Wael, do you understand?" When I said yes, he responded, "Thank God—then I'm confident that everyone else has understood as well." That was one of the reasons I hated the educational system in Egypt. I was very defensive and believed that it was the system, not me, that was blocking my progress. Yet even though I was losing at school, I was winning somewhere else.

Earlier, during the summer of my preparatory year at the university in 1998, I had created a website to help Muslims network with one another. It was pretty much like a simple version of YouTube. There were three fundamental differences, however: it was a website for audio material, not video, since video quality was not as advanced as it is today; content uploading was restricted to me and a schoolmate, since the content was religious in nature; and, finally, the website administrators had to remain anonymous. The webmaster could be reached only via an e-mail address that did not include his real name. I named the website IslamWay.com.

State Security would have immediately targeted me if it had discovered that I was the creator of an Islamic website, no matter how moderate it might have been. When I received the call from Captain Rafaat, I prayed that it would have nothing to do with my IslamWay days. Luckily, he never mentioned it during the interrogation, so I didn't either.

It wasn't too long before IslamWay became one of the most popular Islamic destinations on the Internet. During its early years, the website contained more than 20,000 hours of audio recordings of religious sermons, lectures, and recitals of the Holy Qur'an. Over 3,000 hours of this material I had digitized myself. In addition, the website relied on more than eighty volunteers, the true identities of most of whom remain unknown to me to this day, to collect and digitize content from existing cassette tapes.

Two years after the launch, the website had strong traffic from tens of thousands of daily users. I wanted it to serve as a kind of public library featuring a complete range of moderate Islamic opinions. When the English version launched in 1999, it spread strongly among Muslims who did not speak Arabic and among others who wished to learn about the faith. The website was becoming increasingly influential.

Surprisingly, IslamWay led me to my future wife. Despite my young age, I wanted to get married. I had proposed several times to Egyptian girls whom I met online or through my network of family and friends. My proposals were always met with skepticism leading to rejection. Many families thought I was crazy to seek marriage while I was still at school, despite the fact that I was financially independent and making a decent income. Stubborn and independent-minded as ever, however, I was determined to solve my problems my own way. Somehow I settled on a solution: I decided that what I really needed was to marry a non-Egyptian who would convert to Islam. I admired the openness of American culture and the practical way in which Americans faced life's problems—so not just any Muslim convert, an American Muslim convert. I figured that anyone who changed her faith after a period of contemplation must be someone special—in today's hectic world, most people barely have enough time to think about the ideologies they inherit from their parents, let alone conduct comparisons with other faiths. And even fewer people, I figured, are actually able to cope with the emotional baggage that family and society throw at a person who changes her faith. There was only one difficulty: I did not know a single woman who fit this description. But I did know how I could find one: the Internet.

I first met the woman I was to marry online after reading something she wrote on the website's discussion forum dedicated to new Muslims, where she participated frequently as she practiced the faith she had recently embraced. I reached out to her, and we began corresponding. I found her personality strong and her writing style quite appealing. Yet when I made the crazy suggestion that she visit Cairo—she lived in California—she refused. Our correspondence trailed off over time.

Not too long after, in June 2001, when I was twenty, I planned a trip to the United States in order to donate the website to a U.S.-based charity that supported Muslim communities around the globe. The site had become very successful, and it was now so large that it was beyond my capacity to keep up with its growth. I was working at least thirty hours per week, and my studies were suffering. I had received an offer in 2000, from a close friend who knew I was the owner, to buy 10 percent for $100,000. It was a huge sum for a young man, but I refused to sell. I had never intended to make money from the portal—I do not feel comfortable profiting from social activities. I always knew I wanted to donate it to a charitable organization. Now it was time to transform IslamWay into a professionally managed website, and an American Muslim charity was ready and willing to take it on. So I hopped on a plane.

During my stay, an American friend offered to introduce me to a girl whom his wife knew was looking for a Muslim husband. Fate stepped in: she was the very girl I had chatted with online for months. Weeks later, Ilka and I were married.

I did not tell my parents in advance. My mother, I knew, was especially opposed to the notion of marrying a foreigner with a culture different from ours. Two days after the wedding (attended only by my mother-in-law, two witnesses, and an imam), I called my father. To my surprise, he only scolded me in calm tones for not consulting him and my mother. I asked if he could help me by sending a few thousand dollars until I got settled, and he agreed. I asked him not to tell my mother until I found a way to break the news to her as gently as possible. But he must have thought twice about that idea. Minutes later, my mother called and unleashed her wrath at my unilateral decision. She refused to speak to me for months afterward. I would call and call, and she would hang up as soon as she heard my voice. I wrote letters, trying to appeal to her love for me. I expressed how much I loved her. I praised Ilka as gently and insistently as I could manage, stressing her good manners and other great qualities. Nothing worked.

My stay in America left a major and lasting impression. Like any Egyptian who visits the West, I was in awe of the quality of education, the respect for citizens' rights, and the democratic process that gave people voices and allowed them to be active players in the political process. Admittedly, at my young age, I was easily impressed. I drew a conclusion that I repeated to Egyptian friends many times: "We're being fooled in Egypt!" The thing that impressed me the most was the freedom of religious practice – the respect for religions and every human being's right to practice his or her faith. There were many organizations that defended Muslims and their rights. They were free to criticize the American government's policies without fear of any secret police.

Yet not everything was in favor of the United States in the comparison with Egypt. I sensed an individualism in the air that contrasted greatly with my experience back home. In Egypt, a lot of emphasis is placed on the family and on groups in general, which creates an atmosphere that engenders a sort of emotional warmth in spite of its occasional restrictiveness. On the contrary, in the States I noticed that people were on their own in many situations in which they would have enjoyed much social support if they were in Egypt. My brain was in the United States, but my heart was definitely in Egypt.

My initial plan was to stay in the States to finish my degree, because I was so impressed with American higher education. Yet I had a change of heart after 9/11. I will never forget that day. My wife and I were home, and I had woken up early and started working on my computer when, on a discussion board, I found people asking each other to turn on the TV right away. I watched flames emanating from the first World Trade Center building; we all thought at the time that a plane had accidentally crashed into the tower. I woke up Ilka to join me, and shortly after, we both screamed in horror as a second plane crashed into the other tower. I had never imagined that people who claimed to be Muslims could commit such an atrocity. The faith in which I had been raised both unequivocally prohibits the killing of innocent civilians under any circumstances and completely forbids suicide. So I was dumbfounded when I heard speculations in the media that the culprits were Muslims. Over the years I had observed various Western media outlets magnifying the acts of some crazy fanatics and portraying them as representative of Islam. If 9/11 had anything to do with Muslims, I thought, then those who had planned this monstrous murder of thousands of innocent civilians must have been thinking solely about their political ideologies and could not possibly have considered the damage they would do to the image of Islam and Muslims living in America. Or perhaps they couldn't care less.

It wasn't easy being an Egyptian Muslim in America during the weeks immediately following the attack. It sometimes almost felt as if my fellow Muslims and I were personally accused of this atrocious crime. In public spaces, I was keenly aware of every look of suspicion that came my way. Many of my Muslim friends suffered acts of discrimination, including brief arrests and harassment at airports. I was getting tired of being unfairly singled out and had little hope of finding a job, so I began to seriously consider returning to Egypt.

Ilka, of course, was quite attached to her home country, although she too felt alienated by the barrage of criticism of our religion that washed through many media outlets. The fact that she wore a headscarf made her conspicuously Muslim, and this made a woman's life harder at the time. Still, she hesitated for a long time before agreeing to move to Egypt. She had left the United States only once before, on a short tourist trip to Mexico. I remember her saying to me, "I asked some friends online about Cairo, and they said the streets were filthy."

"Yes, I must admit, some streets are dirty, but people's hearts are clean."

The Egyptian people are among the best-hearted and most humorous in the world. They laugh during the darkest of times and find humor in the midst of suffering. Not even sixty years of a regime of fear could change that.

After a heavy dose of persuasion, Ilka agreed, and we flew to Egypt in December 2001, three months after 9/11. I was adamant that we see my mother immediately upon our arrival. Walking into her house right after fourteen hours of flying was actually quite an experience. She was trying to hide her emotions but failed miserably. She didn't even smile when I said hello, and when I introduced Ilka, she offered a cold greeting. Obviously she felt betrayed. Nonetheless, over time my mother could not help warming to Ilka, and she grew to love her.

Shortly after I returned to Egypt I resumed classes, but I also began searching for a job. An old friend of mine, AbdulRahman Meheilba, along with his partner, Ramy Mamdouh, was working with an Internet startup that provided e-mail services to corporate clients and individuals. Gawab.com quickly spread across the Arab world because its e-mail service supported Arabic and it offered 15 megabytes of storage space at a time when Hotmail offered only 1MB and Yahoo offered 2MB.

Because of the entrepreneurial skills I had acquired during my experience with IslamWay.com, AbdulRahman offered me a job overseeing marketing and sales. Without a moment's hesitation I accepted. We worked hard to spread Gawab's services further in the

Arab world. Eventually we managed to reach two million users and secure sustainable revenue by selling advertisements as well as hosting e-mail solutions for businesses and other websites. As Gawab.com grew, so did my paycheck. I became responsible for a team of twelve employees who dealt with clients in different parts of the Arab world. It was fun doing business with people you never met, thanks to the Internet. The growth of the company was exciting, and so was a six-figure offer of a buyout pitched by an Arab investor.

Working at Gawab gave me my first real sense of professional responsibility. Anything related to marketing and sales came to me. I was even responsible for accounting and cash management. It seemed everything was happening at once: in addition to spending long days at Gawab and many hours studying during my final two years at the university, I had become a father: Ilka and I were blessed with a baby girl in January 2003. We argued about who would choose the name; Ilka strategically allied with my mother and eventually got her way. We gave our beautiful little girl the name Isra.

I was ecstatic about being a father. It was strange for everyone else at school, since none of them had children. In general, many colleagues found me quite strange. Some saw that I rushed into decisions and actions without fully contemplating the consequences. They were right. It is in my blood. And not just that: I have always wanted to swim against the current.

Time quickly went by; Isra turned one, and I officially became a computer engineer in June 2004. Because I was a father, I felt even more responsibility to excel, in order to provide for my family. I scored my highest grades during the last year of school, yet my overall grade of 64 percent was "unsatisfactory."

During my work at Gawab and a few months after graduation, I decided to study for an MBA. My job put me in charge of the company's sales and marketing, and I realized how much knowledge I needed—I could not just read a few books and get up to speed. I needed experienced mentors and a vigorous education in business. My first choice was the American University in Cairo, which has a top-quality MBA program, though it charges high tuition fees. It would mean spending over 60 percent of my annual income on my education. As far as I was concerned, the cost did not matter much, as it was an investment that I trusted would reap returns after a few short years. Yet the university made it clear that I was not a strong candidate. My undergraduate grades were not high enough.

I wrote a long letter to the university explaining the reasons behind my low grades. The general system of education in Egypt was to blame, I claimed. I had missed exams during the first year of electrical engineering, then again during the first half of the third year, when I was in the United States, which had unfairly penalized me. I also explained the distractions of my work and early marriage, and I stressed my attempts to overcome them. One of my dearest university professors, Dr. Ahmed Darwish, who was the Egyptian minister of administrative development at the time, even wrote a letter of recommendation for me.

One of the requirements for acceptance at AUC's MBA program was to score a minimum of 500 points on the GMAT. The director of admissions told me that if I was very serious about my application, I should score higher to compensate for my low grades. She said my score should not fall below 550, the average score of their applicants. I took it as a challenge. After two months of intense preparation I scored a 680, which was very high compared to the scores of my Egyptian peers. A short while later, I was finally accepted. I pledged to the admissions office that I would prove my worth and score the highest grades in all my classes.

Two years and sixteen courses later, I graduated with a 4.0 grade point average, the highest possible. I would start each workday at Gawab, travel from there to the university to attend classes, then spend long hours at the library to study. Achieving straight A's became of the utmost importance to me, even though it would have little effect on my career. Yet I did it. My self-confidence was redeemed. I proved to myself that I was not a failure. Ilka was supportive above and beyond the call of duty and stood behind me throughout. She knew that it was my own personal challenge, and despite the fact that I spent little time with her and our daughter, she always encouraged me to keep studying and focusing on my school projects.

The experience of the MBA program at AUC was crucial. Learning the science behind marketing was key to my career progress, and later on was vital to my online activism. The combination of marketing and a concentration in finance enabled me to understand how to study market needs, design products that address those needs, and promote them to target audiences. The finance classes introduced me to the world of business; since I came from an engineering background, this taught me a lot about how to run businesses financially. Little did I know that only a few years later, all this piled-up experience would come in quite handy in promoting a product I had never seen myself marketing: democracy and freedom!

In 2005, during the first year of my MBA, I had the startup bug. It was then that I met a major Arab investor in the field of technology—a chance meeting that I attribute solely to divine assistance. Mohamed Rasheed al-Ballaa was an engineer and a major stakeholder in the National Technology Group. Headquartered in Riyadh, Saudi Arabia, NTG is a multinational conglomerate with more than twenty specialized information and communication technology (ICT) businesses in the Middle East and North Africa, Southeast Asia, South Asia, and the United States. To my delight, al-Ballaa was impressed with my web experience and listened to me pitch a cars portal for the Arab world similar to AutoTrader.com. He invited me to visit his company's branch in the United Arab Emirates to discuss the project further.

Instead of offering to finance a startup, however, al-Ballaa offered to hire me. He wanted to expand aggressively into the Internet business in multiple ways, and effectively gave me the opportunity to be part of a larger team with a mission to help what he believed would change the face of the Arab web. The offer was quite enticing, since my entrepreneurial dream would be partially fulfilled with very little financial risk and my efforts would ultimately have more impact given al-Ballaa's ambitious vision.

After I started working at NTG, I found al-Ballaa to be a model Arab investor. He treated me like a son, to the extent that my colleagues started to refer to me jokingly as Wael al-Ballaa.

One of the companies established by NTG was Mubasher, a Middle Eastern version of Reuters and Bloomberg financial solutions, which provided Arab stock market investors with web-based screens displaying real-time prices to facilitate buying and selling decisions. Al-Ballaa knew the importance of research and economic news for investors. He asked me to take the helm in starting a Cairo-based company to support Mubasher with data and analysis. I was not yet twenty-five.

At the time I knew nothing about the media, or publishing, or even the stock market. I was so ignorant that I did not even know the difference between a stock and a bond. I began avidly reading everything I could get my hands on about stock markets, and decided to take finance-related courses in my MBA. The company was established, and I quickly assembled a team. Needless to say, I was committed to the Internet generation. Most team members were fresh graduates in business and mass communications; I much preferred them to the experienced researchers, who were nowhere near as digitally engaged.

Mohamed al-Ballaa was extremely supportive of the project's launch, both financially and morally. When Mubasher.info was launched, it quickly became one of the main destinations for many small Arab investors seeking information and news on listed companies.

This experience had a profound effect on me. I was heading a large team of more than 120 people. We were always seeking ways to develop and innovate. I tried every possible means to motivate the team to give the company their absolute best. I constantly urged all the employees to improve their own skills and the work environment. The site had more than one million visitors per month, and its reputation spread around the globe. We launched an English version. Several international stock market websites and companies began using our portal as one of their reliable sources for news and information. As a result, the company became interested in expanding and increasing its investments in Egypt.

Mohamed al-Ballaa was in his fifties, but he had the energy of a man in his twenties. He enjoyed taking calculated risks and venturing into areas where others would seldom go. He often told me that the rapidly evolving world around us makes it difficult to formulate five-year and ten-year plans, especially in a tech-related industry. As an investor, al-Ballaa committed his capital to dozens of ventures at once, a spray-and-pray strategy, hoping that just one would strike gold and make up for losses in the other investments. This made a profound impression on me.

The Internet has been instrumental in shaping my experiences as well as my character. It was through the Internet that I was able to enter the world of communications (when I was barely eighteen) and network with hundreds of young people from my generation everywhere around the world. Like everyone else, I enjoyed spending long hours in front of a screen on chat programs. I built a network of virtual relations with people, most of whom I never met in person, not even once.

I find virtual life in cyberspace quite appealing. I prefer it to being visible in public life. It is quite convenient to conceal your identity and write whatever you please in whatever way you choose. You can even choose whom to speak to and to end the conversation at any moment you like. I am not a "people person" in the typical sense, meaning that I'd rather communicate with people online than spend a lot of time visiting them or going out to places in a group. I much prefer using e-mail to using the telephone. In short, I am a real-life introvert yet an Internet extrovert.

My addiction to the Internet made joining Google a dream that I fervently aspired to make a reality. I had heard a lot about how cool it was to work there. The company's founders, Sergey Brin and Larry Page, became among the most influential people in the world after they developed the web's best search engine. Employees at Google are among the happiest in the world. Their intellectual skills are respected and their innovation is appreciated. For years I persistently applied online every time there was a vacancy that matched my experience. I used to joke with my wife and friends by saying, "I want to work at Google even if I have to take a job as a tea boy."

One of those attempts was in 2005, only a few weeks after I joined NTG. Google announced a vacancy for a consultant in the Middle East and North Africa. Without hesitation, at barely twenty-five years old, I applied. My résumé was designed to demonstrate immediately that I was crazy about the Internet.

I had doubts about getting this position, because of my age and lack of experience. The job was to develop the company's strategy for the entire Middle East. Yet to my surprise, I received a call from Human Resources inviting me to start the interview process.

Google is unique in everything it does. Human Resources sent me some documents to read before the interviews. The entire process took a few short months. The last, and ninth, interview was with a vice president, who spoke to me from his office in London. He asked about my experience in mergers and acquisitions along with my web experience, and about my understanding of regional web issues. Although I knew that I certainly was underexperienced, I was still hopeful, because I felt that my character and Google were "plug 'n' play." This is why I was devastated when, two weeks later, I found out that I had not been selected. My wife was dumbfounded and could not understand why I didn't get the job.

My desire to join Google only intensified. It was no longer a matter of employment; it was a challenge, and I was stubborn. I particularly did not want to fail at joining a company that I thought embodied who I was as a person.

In 2008 the company announced an opening for a regional head of marketing, to be based at its one-year-old branch in Cairo. It was the perfect chance, now that I had my MBA in marketing and the Mubasher experience. Of course I applied. I volunteered a study, on my own initiative, in which I explained my vision of what the company's strategy in the region should be. My observations included technical notes on the search engine's performance in Arabic. Then a series of interviews began: I met with seven interviewers from different countries and functions. I thought I did very well in most of the conversations. My last interview was with the VP of marketing. I still remember my answer to her question "Why do you want to join Google?": "I want to be actively engaged in changing our region. I believe that the Internet is going to help make that happen, and working for Google is the best way for me to have a role."

After eight months of interviews, I received an offer—I had made it to Google! I was later told by my manager that one of the senior Googlers who interviewed me described me in his evaluation sheet as "persistent and stubborn, just the type a company needs when entering a new market."

My skills and experience were enriched by Google. And I marveled at its culture, which was all about listening to others. Data and statistics ruled over opinions. Most of the time, authority belongs to the owners of information, or as W. Edwards Deming once said, "In God we trust; all others must bring data."

Similarly impressive is the trust Google bestowed on employees, who were empowered to access a lot of internal information that other companies would normally restrict to smaller numbers of employees. Communication and sharing knowledge among employees was key to the company's success.

Google did not rise to the peak of the tech industry by luck. Its success is all based on strategy and philosophy. Attention goes not only to employees but also to users. The company listens to its users, asks their opinions, analyzes their usage behavior, and uses this input to develop its products. Teams within the company are constantly changing and developing products using unique and innovative product development methods.

The culture of experimentation was another thing I loved about the company. An experiment is always welcome, so long as the results come quickly. In a case where there is a difference of opinion about features of a product undergoing development, the product managers and engineers will put a beta version out to a group of users. The decision will then be based on the results and feedback. Google is not afraid of failure. Failure is accepted. If a product fails, it is terminated. Simple.

What attracted me most was Google's 20 percent rule. The company allows employees to work on whatever they please for 20 percent of their time (one day a week). This means that they are free to work on projects other than their official assignments if they want. The idea is based on the notion that people work best when they work on things they are passionate about. A host of Google's most outstanding products were born out of the 20 percent rule, including the e-mail service, Gmail, and the largest online advertising management network, Adsense. For me, Google helped reinforce the idea that employee engagement is the most important strategy of all. The more you can get everyone involved in trying to solve your problems, the more successful you will be. I found it natural, a few years later, to apply this philosophy to political and social activism.

A year before I finally joined Google, when I sat across the desk from Captain Rafaat of State Security, he asked me many questions about my religious faith and practice but none about my Internet experience. After a few hours of interrogation, during which he found nothing to hold against me, the State Security officer seemingly decided that I was not a threat in any way to security or to the political status quo. He said that he would try to remove my name from the airport arrivals watch list after presenting a report to his superiors. I thanked him and departed, grateful that this strange day had come to a peaceful end.

If Captain Rafaat and his colleagues had spent more time thinking about the Internet than classifying Egyptians by type of religious belief, they might have been better prepared for the digital tsunami under way.

2

Searching for a Savior

"I'm not into politics." I used to say this all the time, reflexively, whenever the subject came up. It was a popular stance, shared by most Egyptians. It was the result of a deeply rooted culture of fear. Anyone who dared meddle in politics, in opposition to the ruling National Democratic Party (NDP), took a risk, with little hope of reaping any return. Most of us shied away, believing that we could not do anything to change the status quo.

The truth is, however, Egyptians have always expressed political opinions, but only passively. We complain about education, health care, the economy, unemployment, police brutality, bribery, and corruption, but that is as far as we once dared to go. Few would point fingers at the officials responsible, while most kept such thoughts to themselves.

Egyptians who grew up in the fifties and sixties endured the worst repression in our modern history, including arrests, torture, military trials, and other forms of oppression. Most of them chose safety over activism. Informers were so deeply planted that many Egyptians were afraid to discuss politics in public. This generation raised their children first and foremost to fear politics and State Security. Sometimes it seemed to me that we feared the wrath of the secret police more than we feared death itself.

Egyptians practically never chose a president. The dynasty of Mohamed Ali, who is regarded as the founder of modern Egypt, ruled for almost 150 years until the revolution of July 23, 1952 (in a sense, Mohamed Ali himself was installed by popular demand, when a group of prominent Egyptians insisted in 1805 that the former governor, Ahmad Khurshid Pasha, step down). From 1952 on, the military made all key decisions. The army officers who led a military coup against the ruling monarchy chose Mohamed Naguib as Egypt's first president, transforming the nation into a republic. Two years later the Revolutionary Command Council forced him to step down, and they kept him under house arrest for the short remainder of his life. According to Naguib, this happened because he had planned to hand over control of the country to civilian leadership.

Naguib was succeeded by the extremely charismatic Gamal Abdel Nasser, best known for his pan-Arab nationalism. He was highly esteemed by Egyptians, although a lot of his actions actually planted the seeds of repression and autocracy. Under Nasser, democracy meant referendums on his popularity in which people voted either yes or no, and he somehow always garnered 99.9 percent of the vote. Egyptians joked about tracking down the 0.1 percent that opposed his rule.

Nasser's vice president, Mohamed Anwar al-Sadat, became president when Nasser passed away in 1970, with no help from any electoral process. A referendum confirmed him as president soon after; he received 90 percent of the votes. The same scenario occurred when Sadat appointed Mohamed Hosni Mubarak as vice president. When Sadat was assassinated in 1981, Mubarak took over. Potemkin referendums continued to provide a façade of legitimacy. The percentage of "yes" votes changed slightly over time but always remained in the 90 range:

Gamal Abdel Nasser	1956	99.990%
Anwar al-Sadat	1970	90.040%
Anwar al-Sadat	1976	99.939%

Hosni Mubarak	1981	98.460%
Hosni Mubarak	1987	97.120%
Hosni Mubarak	1993	94.910%
Hosni Mubarak	1999	93.790%

Mubarak ruled for five terms, each of which lasted six years. His best terms were the first and second, when he released political prisoners arrested by Sadat and promised widespread reforms. He vowed to fight corruption. He also pledged not to rule for more than two terms, as the constitution required. Many political analysts believe that Mubarak did not start out as a corrupt man. But Lord Acton's rule prevailed: power corrupts, and absolute power corrupts absolutely.

Mubarak, like the presidents before him, held almost all the reins of power in the nation. There was a parliament to issue laws and in theory to divide power with the executive, but in practice the members were kept closely dependent on the regime. Their loyalty was maintained through what came to be called the "chain of interests": privileges and benefits were showered on any parliament member from the ruling NDP. From land to loans to immunity from arrest to (most important) influence — these members were among the country's movers and shakers — a chain of corruption bound them tightly to the regime.

Councils in each governorate of Egypt were selected in the same manner. Known as the Local Popular Councils, they were responsible for services and policies in their respective governorates. The fortunate members who were loyal to the NDP were akin to Communist Party members in the Soviet Union: they received special privileges unavailable to others.

Little by little these privileges eroded the rule of law. The higher up in the chain you were, the less restricted you were by the law. We suffered chronic inefficiencies because of widespread bribery and corruption. The system eroded the Egyptian character. We lost our self-confidence. The phrase "There's no hope" became customary, especially among young Egyptians. For too many of us, dreams of an apartment, a marriage, and a decent life faded. Out of hopelessness came anger. We were ripe for revolution, even when we were terrified by the idea.

When Mubarak broke his promise of a two-term presidency in 1993, state media—the only media at the time with any effective reach—portrayed him as the epitome of wisdom, the only hope for the nation. The pharaoh's favorite cloak, "stability," was the primary argument advanced by the official press. The president was presented as the only viable alternative to chaos. As the ancient proverb put it, "The people you know are better than the ones you don't."

At the turn of the millennium, and after Mubarak had had four presidential terms, the first son, Gamal Mubarak, began—cautiously—to dip his toes into political waters. Rumors were floated to test reactions to the possibility that Mubarak Junior would become president. In nearby Syria, Bashar al-Assad had succeeded his father. Why not the same for the Mubarak dynasty?

Throughout Mubarak's reign, the most enduring and influential opposition came from the Muslim Brotherhood (MB), formed in 1928. The Brotherhood's popularity was regularly presented to the West as a scarecrow whenever Mubarak was under pressure to reform and democratize the regime. Members of the Brotherhood were widely arrested, subjected to military tribunals, and vilified in the press.

The regime played a typical tyrant's game. It needed a bogeyman, so it both repressed and enabled the Brotherhood. Yet after years of obsession with its chosen enemy, the Mubarak regime may have become complacent about other threats. In 2004 a group of opposition activists founded the Egyptian Movement for Change, otherwise known as Kefaya, which means "enough" in Arabic. Kefaya opposed the renewal of Mubarak's presidency for a fifth term and also rejected the attempt to transfer power to his son. The movement's motto became "No to renewal, no to the inherited presidency." Members of Kefaya were diverse, including dissidents, intellectuals, journalists, Internet bloggers, university students, and artists. It was the first group to openly and explicitly express opposition to Mubarak's presidency as well as to his son's potential candidacy. Its first major protest against the regime was on December 12, 2004 (though many of the protesters knew one another from earlier gatherings to protest Israeli strikes on the Gaza strip and the U.S. invasion of Iraq).

The regime did not crack down on Kefaya as hard as it had on the Muslim Brotherhood. The security masterminds could not imagine such a movement mobilizing significant popular support, since many of its members were intellectuals, whose discourse is not usually appealing to the masses. And the regime was right—Kefaya never achieved a broad following. Yet just by exhibiting the courage to protest, Kefaya helped tear down a psychological barrier. And by criticizing Mubarak openly—the group's famous chant became "Down, down, Hosni Mubarak"—Kefaya members were certainly brave pioneers.

Kefaya's courage, however, meant very little to Mubarak Junior. Gamal Mubarak was born in 1963 and graduated from high school in 1980, the year I was born. He received his bachelor's and master's degrees in business at the American University in Cairo. A few years later he left Egypt to work for the Bank of America in London. With a few colleagues, Mubarak then left Bank of America to set up a London-based private equity fund. Upon his return to Egypt in 1998, his political ambitions started to become more obvious, and he joined his father's party in 2000. As the son of the pharaoh, he rapidly became a key person in the party, which he wanted to restructure and reposition. He was promoted to lead the party's Policies Committee, the most important division of the NDP. In addition, he became the deputy secretary-general. He was the youngest man of any consequence in an aging party.

In 2004, a new cabinet composed of Mubarak Junior's close allies was sworn in. It came to be called the businessmen's government, because most of the ministers were rich businessmen. Mubarak nevertheless left the regime's main pillars intact. The ministers of defense and interior affairs and the head of intelligence remained in their positions. Many Egyptians hoping for real change, including myself, were still pleased to see younger faces in government positions. The new prime minister, Ahmed Nazif, had a solid background in technology. Yet it was clear that the regime intended to groom Gamal Mubarak as the nation's next president.

When Gamal Mubarak appeared on the Egyptian scene, I thought it was an opportunity to empower the younger generation and get rid of the old mentality that had been dragging us into the dark for ages. He seemed like a progressive person who appreciated experience and understood the youth culture better than the dinosaurs around his aging dad. The new campaigns for the party seemed to indicate a real desire for change, but later it became obvious that this was purely cosmetic—a change in the campaign but not in the product itself. Corruption was deeply rooted within the NDP, and it seems that Gamal Mubarak agreed to play by the same rules as everyone else.

The following year, 2005, owing to pressure from the international community, parliamentary elections were held under the supervision of the judiciary for the first time. Gamal Mubarak's influence was growing. He had announced reforms within the ruling party (as head of the Policies Committee). The new cabinet was made up of his own men, not his father's, and the party was coming under his control.

Yet the election's first and second phases (out of three, in different locations) dealt a strong blow to the NDP. The Muslim Brotherhood gained seventy-seven seats, bringing them and other opposition groups close to having a third of Parliament's members. If that proportion continued in the third phase, the opposition would have an effective veto over legislation. The message was clear, and alarming: many Egyptians hated the NDP and would vote for anyone who stood up to its political monopoly. In those first two phases, the state police were nowhere near as aggressive as they had been in previous elections.

In phase three, however, the regime showed its true face, blatantly rigging the results. Hundreds of polling stations were shut, and when voters protested, they were handled aggressively. The international community hardly protested, after witnessing the result of fair elections, since the West was wary of the Muslim Brotherhood, whom many regarded as extremists. More than nine people died during phase three, and the Brotherhood won only eleven seats. The result left the MB as the only strong opposition force in Parliament, with 20 percent of the seats. Despite the fact that official NDP candidates won fewer than 40 percent of the seats, the party ended up with 72 percent representation, since many independent candidates joined the party after winning, either because they desired the personal riches associated with each loyal seat or because they were too afraid to decline, or both. It was very clear that the party needed a monopolizing majority to pass any legislation without having to negotiate with any opposition groups in the country. When the emergency state was up for its biennial renewal, the party wanted at all costs to avoid a vote against it. The regime's chief tool of oppression could not be placed at risk.

The same year also brought yet another staged attempt to polish the regime's image in the eyes of the international community. A presidential referendum was turned into a simulacrum of a competitive presidential election. Practically speaking, only leaders of political parties were allowed to run against Hosni Mubarak. State media at that time continued to promote the regime. Stories were written before the referendum to hail his presidential victory as a historical event: Mubarak would be the first Egyptian president to allow competition within an electoral race for presidency.

To say that the Egyptian opposition parties were weak and fragile is an understatement. They were effectively nonexistent. I always used to say that if all the non-NDP parties had united to form one group, its sum of members and supporters would have barely filled Cairo Stadium's 80,000 seats. The regime had even created a regulatory body that had to approve all potential political parties before they could see the light of day. Ironically, it was headed by the secretary-general of the NDP. It is no wonder that almost no new parties were formed during this era of autocracy.

The 2005 elections were truly comical. One candidate promised to bring back the tarboosh, a cylindrical red hat that men wore until midway through the last century, if elected. Another candidate proclaimed that he personally would vote for Mubarak as the man most qualified for the job.

Gamal Mubarak played a prominent role in the 2005 presidential campaign, and his father appeared in public for the first time ever without his regular formal attire. He wore a tieless shirt in an attempt to look young and energetic, although he was seventy-five years old. (He had always dyed his hair black to look young, but this was a bigger change.) In addition to glowing coverage in the state's media outlets, positive PR proliferated thanks to businessmen and shop and café owners upon direct orders from the security apparatus in different parts of the country.

Employees of the government and public sector, who amount to more than six million Egyptians, were given orders to vote for President Mubarak. The final tally was ludicrous: 88.6 percent for Mubarak. Mubarak then cracked down on the two true opponents. One was Ayman Nour, head of Al-Ghad ("Tomorrow") Party. Nour was sentenced to five years on allegations of fraud. Similarly, Noman Gumaa, head of Al-Wafd Party, was removed from his position and expelled from the party's headquarters. If you ran against Mubarak and you really meant it, you suffered.

We all knew it was a sham. The question was, would we put up with it?

Egypt's economy continued to suffer despite the new cabinet's optimistic promises. The regime had been selling off state-owned companies since the 1990s, in an attempt to privatize and vitalize major sectors of the economy. Yet the public was convinced that those deals had been corrupt, and in practice economic conditions had not improved. As a result of their incessant suffering, workers could no longer stay silent. Egypt began to witness a new wave of strikes in 2006 and 2007, in numbers of up to 26,000 protesters at a time seeking social justice. It became obvious that a snowball was gradually forming.

In 2008 workers at Al-Mahalla Textiles called a strike on April 6. This time, Internet activists decided to support the strike, following a suggestion made by a prominent dissident to spread it to all of Egypt. One of the strike's Facebook pages attracted over 70,000 members—this at a time when most opposition demonstrations attracted barely a few hundred protesters.

Several forces helped make the April 6 strike a popular one, if not enormously so. Many groups promoted it, including Kefaya, the two opposition parties (Al-Wasat Party and Al-Karama Party), and several professional associations (the Movement of Real Estate Taxes Employees, the Lawyers' Syndicate, the March 9 Movement of university professors, and the Education Sector Administrators' Movement), not to mention the youth movement that had emerged online for the first time. Members of the latter group came to call themselves the April 6 Youth Movement. It was a loose coalition of many small groups.

Many Egyptians who feared protests and potential arrests found it easier to accept striking. All they had to do was skip work rather than face security forces. Yet many people were disappointed by the strike's minimal results. There was no discernible impact on Cairo's streets or in other big cities. Personally, I noticed some limited street activity on that day. I did not join the strike, as I was not politically active at the time, although I was happy that some Egyptians were finally speaking up for their rights. In the Mahalla, on the other hand, two worker activists were killed, and the city briefly turned into a war zone between workers and security forces. A large outdoor poster of Mubarak was pulled down and kicked by protesters. A video of this historical moment was posted on YouTube, but of course such images could never be seen in mainstream media.

Minimal or not, April 6 sent out a clear signal to everyone that the Internet could be a new force in Egyptian politics. The security force's reaction was to develop a new division dedicated to policing the Internet. Similarly, the NDP established an "Electronic Committee" rumored to have legions of well-paid young men and women whose mission was to influence online opinion in favor of the party through contributions to websites, blogs, news sites, and social networks. Arrest orders were issued for April 6 activists, and they became fugitives. The young activist Israa Abdel Fattah was arrested on the day of the strike because she founded the largest Facebook group promoting the strike online. She was released a little over two weeks later.

I resented the regime more than ever but still wondered what I could do about it. I was not optimistic about the impact of the activists' efforts, and I was also busy with work, where I spent all my time. Nevertheless, I was inspired by the courage of those heroes who stood up to the regime at the height of its strength. They risked their lives for the dream of change. The Egyptian revolution will remain indebted to everyone who tossed a stone into the still waters at a time when doing so risked beating and arrest, or worse.

One of the April 6 Youth Movement's prominent young figures, Ahmed Maher, was chased by the police a few weeks after the strike. He tried to escape by car, but he was caught, beaten badly, and dragged to a State Security branch, where he was brutally tortured. Security forces were in disbelief: how had opposition youth groups emerged without any political affiliations, Islamist or other? They fell back on their usual strategy: set an example with group leaders, so that other dissidents would think twice before joining their movements.

Ahmed Maher was released days after his abduction. He headed straight to a human rights activist, who took pictures of his tortured body. Like other audacious young men, Ahmed refused to back down. He went to the media, seeking the protection of public opinion. He was right: regimes of terror cannot stand exposure.

And increasingly, technology made public exposure inevitable. Egypt has seen a significant shift in media patterns over the past decade, thanks to the rise of privately owned printed newspapers and magazines and the spread of satellite television. The private media are not as tightly controlled as the official state-owned media, but they have faced their share of manipulation. Many famous anchors and talk-show hosts have been forced out of their jobs. Still, the new private outlets have produced more even-handed stories, even though their owners tend to have strong connections to the regime.

The Al Jazeera satellite TV channel, established in 1996, also played a significant role. The channel's talk shows offered heavy criticism of many Arab leaders. Within a few short years, Al Jazeera became the most viewed channel in Egypt and the entire Arab region. The network set an example that has been followed by many channels throughout the Middle East.

In parallel, the number of Internet users in Egypt increased rapidly, from a mere 1.5 million in 2004 to more than 13.6 million by 2008. Discussion forums, chat rooms, and blogs flourished, providing an outlet for many users to express opinions freely for the first time. State Security occasionally arrested and harassed bloggers for discussing sensitive issues and for sharing news that the regime didn't like. Yet the number of politically focused bloggers only increased.

In the early part of the decade, I was only passively opposed to the regime, like many of my countrymen. I regularly read the opinions of the most daring opposition columnists, such as Ibrahim Eissa and Fahmy Howaidy. I closely followed the Muslim Brotherhood's website to remain up-to-date with their news. At most, from time to time I initiated political satire of my own, anonymously circulating jokes on the Internet.

One of my jokes, in 2003, was an image satirically depicting President Mubarak's Hotmail in-box. The unread e-mail included a message from President George Bush with the subject line "Mubarak, how can I be president for life?" Another e-mail, from his son Gamal, asked if he could inherit the presidency as Bashar al-Assad had; another was a Swiss bank statement declaring the president's balance to be $35 million. The trash icon in this design carried the title "The People's Demands." This image spread like wildfire, but I carefully kept from claiming credit.

I expressed my opinion of the regime only to friends and family, and they always warned that I was asking for trouble. When the debates got heated and I was eventually asked, "So what's the alternative?" I could only say, "Any alternative would be better than this regime." Most people did not find this answer convincing.

• • •

The absence of alternatives was a key part of the oppressor's master plan. Any popular figure who surfaced, presenting the remote possibility of an alternative to Mubarak's iron rule, was swiftly denounced, defamed, or eliminated. It had happened to the former minister of defense, Mohamed Abu Ghazala, former prime minister Kamal al-Ganzoury, and the former minister of foreign affairs, Amr Moussa. A lot of Egyptians thought that these men had been forced to resign from their posts and retreat from public life because of their popularity on the street. I couldn't agree more; Mubarak was so paranoid that anyone he perceived as competent became a threat to him.

We all craved an alternative. We needed a savior, and we were ready to pour our hopes onto any reasonable candidate. Finally, two years after the April 6 movement began, Egyptian activists believed they had found one.

Mohamed Mostafa ElBaradei, the former chief of the United Nations' nuclear watchdog, the International Atomic Energy Agency (IAEA), is a patriotic Egyptian who had worked in politics since he received his law degree in 1962. ElBaradei showed great skill as a diplomat. His diplomatic career began in 1964 in the Ministry of Foreign Affairs, with postings at the United Nations in New York and in Geneva. He rose to become special assistant to the foreign minister in 1974. He earned a law degree from New York University and then returned to the foreign ministry until 1984, when he became a legal adviser to the IAEA. In 1997 he became its director-general. ElBaradei and the IAEA received a Nobel Peace Prize in 2005 in recognition of efforts to limit the proliferation of nuclear weapons in the world.

Here was a man whom the Egyptian regime could neither eliminate nor easily tarnish. At first the regime tried to embrace him. State media hailed Dr. ElBaradei as a source of national pride. He was granted the highest state honor, the Order of the Nile, by President Mubarak in 2006. As the fourth Egyptian ever to win a Nobel Prize, he was lionized on the street.

In 2009, as his third term as director-general of the IAEA was nearing an end and he prepared to return home, ElBaradei told Egyptian newspapers that he was unhappy with the way Egypt was governed. He focused his criticism on the lack of democracy and the low levels of public health and education. Not surprisingly, he disappeared from our state-owned media.

Nonetheless, his popularity could not be easily snuffed out, thanks in part to Internet activists. A university student by the name of Mahmoud al-Hetta decided to start a Facebook group called "ElBaradei President of Egypt 2011." ElBaradei had been asked by CNN if he would run for office, and he had replied that it was premature to answer such a question. That indefinite reply left the door open, and young Egyptians began mobilizing support for him.

Many of the young people aspiring for real change in Egypt joined the Facebook group. Finally we had an answer to the question "If we don't vote for Hosni Mubarak, who will we vote for?" Tens of thousands of users joined the Facebook group, and among them I recognized many personal friends who hitherto, like myself, had never been involved in politics. We all saw a glimmer of hope for reforming Egypt. Mahmoud al-Hetta and others used spontaneous online methods to invite ElBaradei to nominate himself for president. Shortly thereafter, the group's popularity crossed the 100,000-member mark and ElBaradei announced his desire to play an active role in Egypt's movement for change. His wish was for Egypt to reclaim its historic status and become a true democracy, not just a nominal one.

The Egyptian regime was taken by surprise and lost its balance. Instantly the powers-that-were launched a defamation campaign. The man who most represented our national pride was suddenly subject to a series of false accusations. In record time he was depicted as an ally of the United States, with a Western agenda, and even portrayed as the main reason for the United States' war on Iraq, which resulted in hundreds of thousands of dead Iraqis (ElBaradei is known for his opposition to the war on Iraq and his attempts to contain crises through diplomatic means rather than bloodshed). He was said to be a glutton for power; after three terms at the IAEA he now wanted to rule Egypt. Proponents of the regime claimed that he lacked political experience. They even started a rumor that he held Austrian citizenship (since he had lived in Vienna for many years). The absurdity reached its peak when the chief editor of the nation's largest newspaper claimed that ElBaradei had been a failure as a schoolboy and that his grades were the worst in his class during one of his elementary school years.

The Facebook group was undaunted by these transparently absurd charges. The poet Abdel Rahman Youssef emerged as the campaign manager for the movement created by the group in an effort to try to venture out into the field. In December 2009 the independent newspaper *Al-Shorouk* published a long interview with ElBaradei, two months before his planned return from the IAEA. Over a span of three days, he told the reporter about his ambitions for change in his country.

The highlight of that interview was ElBaradei's conviction that change was inevitable in Egypt. He added that he refused to run for president in a sham election, if the regime was to exploit lifeless political parties again to project Mubarak as the country's only option. ElBaradei's refusal to grant legitimacy to the regime was his first confrontational step. Yet the constitutional amendments that he urged as necessary before he could run for president were perceived as far-fetched by most people.

As the political scene in Egypt was changing, so was my personal life. Ilka was getting very frustrated with life in Egypt. She found it impossible to drive in Cairo's jungle of a traffic system and simply could not adapt to the pollution. She struggled with Arabic as well, and had trouble managing day-to-day activities. For these reasons, among many others, she was not happy living in Egypt, even after seven years of residency there, and my regular absence from home only served to reinforce her feelings of alienation. At the same time, Google's Middle East team was beginning to centralize in the UAE, and it was gradually becoming more convenient for my career to move there. When I consulted Ilka, she was strongly in favor of a move. I was quite hesitant, as I preferred to stay in Egypt, yet it was becoming clear to me that this would be a selfish thing to do. Finally, in January 2010, I relocated to Google's office in Dubai, but fortunately the nature of my work would take me to Cairo on a regular basis. Ilka was thrilled to be in Dubai, and I must say that I enjoyed it as well, although my heart remained in Egypt.

I continued to follow the heated debates back home closely. I accessed the Facebook group on a daily basis to read the discussions, but I was not yet actively involved.

Dr. ElBaradei's return to Egypt was scheduled for February 2010. Many of the country's political forces organized a reception for him at the airport, in the form of a few hundred activists who were willing to face the consequences of publicly opposing the regime. Several Egyptian public figures joined them, including the veteran TV presenter Hamdy Kandil, whose show had been taken off the air because of his outspoken criticism of the regime. What was new, alongside the old opposition guard, was the presence of many young people who ventured out for the first time in support of change.

I was still not ready to make a public statement by attending. I had a lot to lose. My employer, Google, was a dream company voted often to be the world's best employer. I was responsible for Ilka, Isra, and my son, Adam, who had been born in 2008. I also believed, despite my optimistic outlook, that change in Egypt was a difficult challenge that would take time. But so as not to miss out on the chance to be an active part of the movement, I finally decided to leverage my media, marketing, and Internet experience to help develop what later became Dr. ElBaradei's official Facebook page. My aim was to establish an ongoing communication channel between him and his supporters.

Personally, I have always hated hailing individuals as saviors, and I do not believe in magical solutions. What I do believe is that real change entails a change of policies and methods, not a mere substitution of leaders and individuals. Egypt's salvation, in my opinion, would never come at the hands of a benevolent dictator. I might not have agreed with Dr. ElBaradei on every single issue, yet I did not hesitate to support him as a presidential candidate. My enthusiasm was for the idea rather than the person, but the only way back then was to support an idea through a person. The regime resembled a wall of steel. It had to be weakened little by little. Egyptians needed to be offered alternatives.

The thing I admired most about ElBaradei was his self-perception. He asserted repeatedly that he was not a savior and that the Egyptian people needed to save themselves. He put himself forward as only a tool in support of the cause. To me, he was a professional, well educated, someone who could speak to the ambitions of Egypt's youth.

As an experienced Internet user, I knew that a Facebook page was much more effective in spreading information than a Facebook group. As soon as someone "likes" a page, Facebook considers the person and the page to be "friends." So if the "admin" of the page writes a post on the "wall," it appears on the walls of the page's fans. This is how ideas can spread like viruses. A particular post can appear on the users' walls to be viewed thousands, or even millions, of times. In the case of groups, however, users have to access the group to remain updated; no information is pushed out to them.

So I created a page in February, days before ElBaradei's arrival, and I began its promotional marketing campaign. The number of fans who "liked" the page exponentially increased because of the sheer number of ElBaradei enthusiasts. I updated the page with excerpts from ElBaradei's interviews, and I highlighted his vision for reform in Egypt as well as his emphasis on the country's need for true democracy.

A few days after creating the page, I figured that I needed a coadmin. The nature of my work for Google required me to travel a lot, and I didn't want the page to be dependent on my personal schedule. I noticed that one of the people on my Facebook friends list was also quite excited about ElBaradei. I had never met AbdelRahman Mansour in person, but we had been virtual friends since August 2009. AbdelRahman was a twenty-four-year-old undergraduate finishing his last year of journalism study at Mansoura University, 120 kilometers away from Cairo. His activism began when he started blogging about Egypt's political situation. He had covered the rigging of the 2005 elections, among other crucial events at the time. I found his status updates on Facebook and Twitter to be thought-provoking. At one point, when I sent out an open invitation to all my friends to join the page, I received a message from AbdelRahman asking if I was the admin behind it. He instantly became an appropriate choice for a co-admin. On the one hand, I admired his enthusiasm and intellect, and on the other hand, he had now become one of the very few people who either knew or suspected that I had founded ElBaradei's Facebook page. Without hesitation, AbdelRahman accepted my offer. That day would mark the beginning of a virtual working relationship that still continues today.

Naturally, it took some time to build mutual trust and understanding. Many times I would send private messages asking AbdelRahman to remove content that he posted on the page, and we would occasionally have heated discussions about such matters. Whenever push came to shove, however, I had the final say. The golden rule was to ask ourselves the following question: "Would Mohamed ElBaradei write this post himself?" This made our decision-making process much easier.

Soon after his arrival, ElBaradei met with key opposition figures. Immediately following the meeting, we were surprised to receive an announcement of the establishment of a newly formed body called the National Association for Change. The idea was to bring together everyone known to oppose the Egyptian regime. Members included the former presidential candidate Ayman Nour; the media veteran Hamdy Kandil; Dr. Mohamed Ghoneim; some leaders of the Muslim Brotherhood, such as Muhamed el-Beltagy, a former MP; some political parties, like the Democratic Front, Al-Karama, and Al-Wasat parties; the Revolutionary Socialists; Egyptian Women for Change; the April 6 Youth Movement, and others. The association's first action was to release a statement entitled "Together for Change," or what was also known as "ElBaradei's Seven Demands for Change":

1. Terminating the state of emergency
2. Granting complete supervision of elections to the judiciary

3. Granting domestic and international civil society the right to monitor the elections
4. Granting equal time in the media for all candidates running for office
5. Granting expatriate Egyptians the right and ability to vote
6. Guaranteeing the right to run for president without arbitrary restrictions, and setting a two-term limit
7. Voting with the national identity card.

It was an ambitious list. It meant freeing the press; it would enfranchise eight million expatriate Egyptians; and it would help create an independent judiciary, among other spectacular achievements. The seventh demand was crucial for fair elections. The standard voting practice in Egypt was that voters were issued "electoral cards" in their respective districts. The card was required at the polling station for a voter to cast his or her vote. Since rigging was significant and consistent, most Egyptians were disinclined to obtain a card. In turn, that made rigging even easier. As a popular joke put it, we were so proud of our democracy that we even let deceased people cast votes. To demand that voting require only a national identity card was to demand free and fair elections.

The great thing about these demands was that the majority of opposition forces agreed to and supported them. Even the regime found it difficult to argue publicly against most of the seven demands. Dr. ElBaradei's idea to issue this statement as a petition was a great one. It was an excellent new tool of pressure, and it increased the possibilities that the regime might compromise.

To collect signatures in significant numbers, the movement turned to the Internet. The petition was published online, and citizens just needed to enter their name, address, and national ID number to sign. The organizers also helped people overcome their fear by publishing the initial hundred signees, who were public figures willing to use their authentic personal information.

Fear overcame me on the first and second days of the petition. But then I entered all my personal information and signed. I was citizen number 368 to do so. My fear turned into excitement when I realized I was beginning a new phase: I now publicly opposed the regime. I had no doubt that State Security downloaded the list of signees regularly, particularly since it contained everyone's full name, yet I was excited to be part of the growing crowd.

I was keen to meet Dr. Mohamed ElBaradei, and I tried to schedule a meeting with him during my first trip back home. I sent an e-mail to the Egyptian actor Khaled Abol Naga, whom I had first met at a Google event that we organized for Orphans Day in April 2009. I had seen him endorse ElBaradei on YouTube. I explained that I wished to augment ElBaradei's efforts with my Internet abilities. Abol Naga's response came instantly, providing the e-mail address for Ali ElBaradei, Dr. Mohamed ElBaradei's brother.

I e-mailed Ali ElBaradei, introducing myself and explaining that I managed the ElBaradei Facebook page. He did not know about the page, yet he welcomed any kind of cooperation and promised to set up an appointment with Dr. ElBaradei when I was next in Cairo.

At the same time I e-mailed Mahmoud al-Hetta, who managed the "ElBaradei President of Egypt 2011" group. We spoke on Skype when I was in Dubai and discussed how we could cooperate. I was amazed at how brave this young man was, as were the other activists who used their real names on the Internet. Yet I advised him to hide his name, as Facebook enables you to do, for the sake of the campaign's sustainability. There was no need to publish names where State Security might see them, I said. It was a brief call, and we agreed to meet up as soon as I was in Cairo.

A couple of weeks later, on my way to meeting Mahmoud, I was paranoid. I remained afraid of State Security. When I arrived at the local café on a small side street where we had agreed to meet, I glanced left and right before I joined both Mahmoud and Abdel Rahman Youssef, the campaign manager for the movement on the ground. The poet sensed my apprehension and tried to reassure me. He argued that our work was for a just cause, and that accordingly we had nothing to hide or be afraid of. I was not convinced, and I argued back that secrecy could never harm us and might even prove beneficial to our battle for democracy at a later point in time. I also requested that both of them keep my identity concealed. We discussed the importance of breaking the psychological barrier of fear and how to campaign for the petition with the seven demands. Signees had barely reached 10,000 at the time, a number that fell significantly below our expectations. Although no clear action plan was born out of our meeting, I was nevertheless thrilled to see such zeal and enthusiasm for the cause.

On April 11, I finally had a chance to meet with Dr. ElBaradei himself. His brother informed me by e-mail of the appointment, mentioning that others would also attend. I asked him if I could invite two other people to join us; he didn't mind. AbdelRahman Mansour couldn't make it, as he was out of the country, so I called two other friends who were equally devoted to helping to change Egypt: an engineer, Mostafa Abu Gamra, who owns a technology company that works in content development, and Dr. Hazem Abdel Azim, a senior government official working at the Ministry of Communications. I was quite excited to meet the man whom I had been independently campaigning for.

ElBaradei lives in a villa in one of the private residential compounds on the Cairo–Alexandria Desert Road. I planned to take a taxi, to avoid any potential trouble of being recognized by State Security informants via my car's license plates. Dr. Abdel Azim, however, decided to drive and offered me a ride. ElBaradei was a prominent Egyptian figure and there should be no problem visiting him, he assured me. We met Mostafa Abu Gamra on the way, and the three of us headed off. The guards at the compound's gates let us in without any problem.

The villa was beautifully furnished and decorated, yet it was not extravagant in any way. Some of Dr. ElBaradei's critics claimed he lived a lavish suburban life disconnected from that of ordinary Egyptians. They had portrayed his home as a palace or fortress, with high fences, but this was not the case.

ElBaradei received us in person. Everything he said lived up to my expectations. I was worried that this might change once I offered some criticism; people's true faces appear under criticism, not under praise. He stood among a group of his guests, which included two young film directors, some senior businessmen, and other prominent figures.

Everyone was involved in a heated debate. ElBaradei was an excellent listener, and it never felt like he was leading the discussion. On the contrary, he seemed to be seriously learning from the opinions of others—just the type of leader I felt Egyptians needed. Then I offered my criticism: I suggested that he needed to speak in a language closer to the hearts of mainstream Egyptians. The jargon of elitist intellectuals would not help our quest for popular support.

I also mentioned ElBaradei's recently initiated Twitter account. It was new at the time, but he already had 10,000 followers. It took very little time for him to become the most followed Egyptian on Twitter. I suggested that he sometimes seemed too rushed in his posts. Some of his tweets did not sit well with activists and newspaper readers (newspapers regularly published his tweets). His great quality, if you asked me, was that he refused to be considered a savior. He believed in the nation's youth and in their ability to bring change. I recommended that he tweet about that more frequently. Young Egyptians needed to regain their self-confidence before they could take action.

I also criticized his travels outside of Egypt during these difficult times. Many others viewed this as his worst error. Regardless of the fact that he actually had many scheduled commitments abroad, ElBaradei's frequent travels hurt the perceived effectiveness of the campaign and gave his opponents a chance to taint him as a tool of the West, or a self-promoter who ignored his homeland.

Everyone had something to say. The two directors, Amr Salama and Mohamed Diab, thought that the seven-demands petition was inviting trouble for ElBaradei. Making it a priority and making the signees' information publicly available at a time when dissident Egyptians were not yet ready to go public was not right, they claimed. They had a point: a vast gulf separated the total number of potential supporters and the actual signees up to that day.

On that question, however, I defended Dr. ElBaradei's vision. I found the statement to be an excellent manifestation of the snowball effect. The daily increase in signatures, I believed, made people hopeful. It also prompted community discussions about the statement's seven demands, adding pressure on the government to implement them.

It was a fruitful meeting that left me both optimistic and energized. I took a picture with ElBaradei and made it the profile image on my Facebook page. The caption under it said, "I am Wael Ghonim. I declare my support of Dr. ElBaradei." The meeting had helped me partially break my own barrier of fear.

Next I created a Google e-mail group called "ElBaradei" to enable key supporters to communicate effectively. It was a closed group that could be joined only with permission from one of the moderators. I began adding people whom I knew and trusted to the group. Ali ElBaradei forwarded the e-mail addresses of his brother's other supporters, those whom he thought would add value to the group. Discussions proliferated through this e-mail group, but fieldwork remained limited.

On ElBaradei's Facebook fan page, both AbdelRahman and I tried hard to improve his public image in spite of the government's vicious defamation campaigns. We searched through state press archives available online and extracted articles that praised ElBaradei's efforts. These articles made the recent defamation look absurd: how could a "despised traitor" be a celebrated hero abroad? I found many pictures of ElBaradei with such world leaders as the American president, the French president, the German chancellor, the king of Saudi Arabia, and others. I deliberately published them to stress the fact that ElBaradei was not simply an "apolitical scientist," as his detractors sought to portray him. AbdelRahman even translated and posted the full transcript of his Nobel Prize acceptance speech, in which ElBaradei affirmed his loyalty and allegiance to both his country and his faith.

The core accusation of the smear campaign was that ElBaradei was responsible for the deaths of hundreds of thousands of Iraqis, having misled the United States into believing that Saddam Hussein secretly harbored weapons of mass destruction in Iraq. We were adamant about proving this to be a blatant lie. I found an online video of the UN Security Council meeting at which ElBaradei presented his report asserting that Iraq was free of any weapons of mass destruction. The report demanded more time for inspections and rejected the military intervention proposed by the United States. I added Arabic subtitles to the video and published it, hoping it would show ElBaradei's innocence regarding allegations that he had somehow facilitated the U.S. war on Iraq.

On April 6, 2010, less than three months later, the number of members of Mohamed ElBaradei's page exceeded 100,000. The April 6 Youth Movement also attempted to celebrate its anniversary on that day by organizing a demonstration, but the attempt failed. The security forces were watchful and well prepared.

Online, AbdelRahman and I were restrained. After all, we were writing on behalf of Dr. Mohamed ElBaradei. Our language was formal. We rarely posted our personal opinions, and we were convinced that the page had to present him in a formal light. Most contributors thought that Dr. ElBaradei was personally managing the page. The experience taught me a lot. I had never before managed a Facebook page.

On April 15, I received an encouraging message from Dr. ElBaradei himself, sent through his son. He wrote: "Spent some time browsing the fanpage today. It is wonderful. Many thanks for a very creative and professional job. Keep it up." I replied, thanking him for the support and telling him that it meant a lot to me. I cc'ed AbdelRahman Mansour in the e-mail thread and introduced him as the page's second admin, who deserved as much recognition as I did for all his efforts.

One of the important activities I initiated on the ElBaradei page was the use of opinion polls to make decisions. Despite the fact that Internet polls are far from scientific, they still offer a good means for testing trends of opinion. Besides, in Egypt, offline opinion polls, carried out through actual interviews, were possible only with a permit from the Ministry of Interior. Needless to say, the ministry had no interest in helping political activists gather information from the public.

I located a good polling site that supported Arabic and subscribed to its services. The first poll I developed aimed to measure the page members' level of satisfaction with Egypt's status quo and to explore why many of them had not signed the seven-demands petition. More than 15,000 participants completed the questionnaire. I aggregated and analyzed the results, then sent a message to the e-mail group as well as to Dr. ElBaradei with many recommendations to help increase public support for the petition.

After I provided these comments, Dr. ElBaradei invited me to meet a group of young men who had been working to promote the petition. First I met Dr. Mostafa al-Nagar, who had succeeded Abdel Rahman Youssef as the general coordinator of the "ElBaradei President of Egypt 2011" campaign. Mostafa came across as a sincere person who had a real desire for change. We became good virtual friends. We chatted online frequently about current events, encouraging each other and sharing disappointments. Mostafa was a dentist and political activist of my age who was quite dynamic, and State Security used the emergency law to arrest and detain him for his political dissidence on more than one occasion. He certainly had an abundance of street smarts, and I was admittedly lacking in that department.

We constantly argued about the role of the Internet in the process of change. He believed that the Internet was a virtual world with limited impact on reality, while I found it to be the key vehicle to bringing forth the first spark of change. The Internet is not a virtual world inhabited by avatars. It is a means of communication that offers people in the physical world a method to organize, act, and promote ideas and awareness. The Internet was going to change politics in Egypt, I wrote on Facebook and Twitter, and the 2011 elections would not be similar to those in 2005.

I will never forget the cynical remarks I received in response. A friend joked that the Egyptian regime would change the Internet before the Internet managed to change anything. Many actually believed that the regime would censor the Internet if it represented any sort of threat. Egypt would follow the Saudi Arabian example, they thought, where accessible websites are strictly controlled and citizens are unable to visit antigovernment sources. I did not agree. The Egyptian regime needed to be seen as a progressive, welcoming country to the outside world. Its economy depended in part on tourism, and the regime cared deeply about its global reputation.

Things were moving quite slowly with Dr. ElBaradei's campaign, and most of my recommendations were not implemented. My frustration increased, particularly as the rate of new signatures dropped. Yet I separated my personal feelings from the Facebook page. There I tried to spread hope. Both AbdelRahman and I followed all of ElBaradei's news stories and his field campaigns. We published photos of his campaign visits to places like Old Cairo and Fayoum, and we continued to write his opinions and track the number of signatures on the statement as well as expose the political situation in the country. Many comments on the page demanded that Dr. ElBaradei take more practical steps on the ground and not limit himself to Facebook and Twitter.

One of the decisive moments for me was meeting Dr. El-Mostafa Hegazy, who owns a strategic consulting firm, in his office. He invited me, Dr. Hazem Abdel Azim, and other activists to talk about change in Egypt. He was against the idea that political change should be personified or reduced to a single person's campaign for presidency. His opinion was that it was critical at this phase to focus on change as a goal in and of itself. He wanted to establish the notion that Egyptians owned their country. It would inspire resistance to injustice and corruption across the board. I remember this meeting vividly. I was arguing that promoting ElBaradei was in essence promoting change. But I also agreed with Dr. Hegazy's opinion—before ever meeting him—that positioning ElBaradei as a savior might end up hurting the real cause. After the meeting, the words "This country is our country!" rang loudly in my ears, and they continue to do so to this very day. I wanted every other Egyptian to shout them out as loudly as they could.

A few days later we received an e-mail, in English, from Dr. Abdel Azim apologizing for not being able to continue with the political campaign for change.

> Dear All,
>
> I am very sorry to inform you that I will not be able to be engaged in any political activity related to our hope for change. My position is extremely sensitive as a senior government official.
>
> Although this is known from the beginning, but there was a miscalculation from my side. I was having a very firm position in the last weeks that I would like to continue in this initiative, and I can and willing to resign from my job any time to be free, and actually I prepared the resignation.
>
> Yet the equation was not that simple and my issue was very highly escalated to the extent that I heard signs of real threats of different sorts, on my well-being and on my family.

It was sad to see Dr. Abdel Azim renouncing the efforts to bring about change in Egypt, but none of us could really blame him. We knew that these threats were very serious. Witnessing this firsthand only amplified my conviction that it was very important to work anonymously as much as possible. I kept contact with Dr. Abdel Azim, and occasionally we would chat online and share our thoughts on current events.

Meanwhile, my frustration at the campaign's pace mounted, and I finally decided to send a message to Dr. ElBaradei through his brother. I expressed dissatisfaction with the progress of his campaign and my hope that he would move faster. The movement for change needed to be more flexible and dynamic. He had greatly raised our aspirations, but now we were hungry for actual change on the ground. I expressed my astonishment that we did not meet regularly and that our communication was limited to messaging through his brother. I mentioned that I spent long hours every day promoting his ideas online and that I thought it would be fruitful if I spent at least an hour a week with him, discussing the campaign's strategy. He responded one day later via e-mail, again through his brother. He said he understood my feelings and explained that we were living under exceptional conditions. He was doing everything he could, in spite of the legal restrictions and media assaults he regularly faced. At the time there was no legal framework for our work together, and therefore he preferred to keep our communications indirect. I saw his point, but I believed that the regime could harm us if it wanted to, without the need for legal justifications. Later I settled for meeting Ali ElBaradei in person to deliver my point of view more thoroughly.

When we met, Ali ElBaradei defended his brother. After all, Dr. ElBaradei had stated from the start that he was not a savior. We, the young people, must work harder to collect signatures for the petition, Ali argued. Although the meeting added no tangible dimension to my overall strategy, I once again felt partially relieved after expressing my opinion.

My performance at Google declined significantly during this period, but my manager was still happy. Before I got busy with ElBaradei's Facebook page, I would sometimes spend up to fifteen consecutive hours a day finishing a project, or finalizing a marketing plan for a new product, or simply brainstorming with fellow employees on new ideas for the region. Understandably, my quarterly performance reports at Google always stated that I needed to improve my work-life balance.

Yet my wife was incredibly supportive. From the very beginning, she had known that she was marrying a workaholic who was addicted to living online. Occasionally she would remind me that I needed to give more time and attention to my family. I tried from time to time to improve, but I must admit that no matter how hard I tried, I would always relapse.

The state's campaign to control ElBaradei's growing popularity became fierce. Security authorities had previously issued orders that banned ElBaradei from appearing on Egyptian media. Now private television channels that had previously besieged him for interviews also kept their distance. Coverage in the print media was not as bad as on television, but ElBaradei's news was now featured a lot less than before. Public opinion fell victim to this campaign, particularly as ElBaradei did not make a habit of refuting baseless allegations. Many Egyptians didn't know about the media ban. The only remaining outlet was the Internet. The Twitter account was his favorite channel on which to vent, even though his follow count did not compare to the number of followers of traditional media outlets.

As the situation reached this dire point, I got an idea, inspired by a popular Google product that had been utilized by election campaigns in other parts of the world. Google Moderator is a tool that gives the user the ability to solicit questions from an unlimited number of other users and subsequently to rank these questions based on popularity votes so that they can be answered accordingly. What a cool way to democratize feedback!

I presented the concept of Google Moderator to Ali ElBaradei and explained what it could mean for the campaign. Using this service to hold an event would reach a vast number of Internet users, the majority of whom would be young people whom the NDP had never communicated with in any genuine way. Dr. ElBaradei welcomed the idea and said he was ready to implement it as soon as he came back from a trip abroad.

The initiative was announced on his Facebook page in mid-May 2010. It was called "Ask ElBaradei." The number of fans on the page had now reached 150,000, of whom more than 2,700 participated. They posted 1,300 questions that received about 60,000 votes. It was an astonishing outpouring. Ironically and in contrast, Gamal Mubarak's team had initiated an Internet dialogue shortly before this and asked interested people to send their questions before attending the event. Of course, it was all scripted in advance and the questions were carefully selected.

I wondered what would happen if President Mubarak were to receive questions from Internet users. Would his aides be able to accept clear and direct questions without the usual politicking and deception? The answer was obvious.

The questions for Dr. ElBaradei were profound. Many of the questions that received the highest number of votes revealed anxiety about the follow-up to the signature-collecting campaign. The most important questions were: How will the signatures collected be useful? What is Plan B, if the regime refuses nonviolent change after we collect a large number of signatures? How can we reach rural parts of Egypt to spread awareness about change? Will you take Egypt toward secular governance? What is your position on the second article of the constitution, which states clearly that the Islamic Sharia is the nation's main source of legislation? What are the priorities of your presidential agenda? Finally, Do you eat kushari? (Kushari, made of rice, lentils, and pasta, is a very affordable and popular daily meal for many Egyptians.) It was clear that many people simply needed reassurance that ElBaradei was "one of us."

Together with other coordinators of ElBaradei's campaign, we filtered the questions and began searching for an interviewer who would address these questions to Dr. ElBaradei. Our search was not easy. Everyone we asked refused to play this role; some attributed it to personal reasons or prior commitments, and others said they were afraid of the consequences. In the end, we decided that the campaign's own Mostafa al-Nagar should be the interviewer. The interview was viewed by more than 100,000 online users.

Dr. ElBaradei tried to remain optimistic in his responses. Instead of appearing frustrated at the limited number of signatures and blaming people's passive attitudes, he spoke about proactivity and the importance of joining forces for the sake of Egypt's future. The man was inspiring in his presentation of a better tomorrow. The regime cannot resist the people's demands for long, he said.

Dr. ElBaradei was blessed with optimism. Every time things seemed dark, he beamed with hope and asserted that change was coming. One famous opposition journalist, known for his sarcasm, commented, "He must know something that we do not." And it turned out that he obviously did.

ElBaradei had it right all along: we did not need a savior; we had to do this ourselves.

3

"*Kullena Khaled Said*"

On June 8, 2010, while browsing on Facebook, I saw a shocking image that a friend of mine had posted on my wall. The picture linked to the official Facebook account of Dr. Ayman Nour, the former presidential candidate who was a political activist. It was a horrifying photo showing the distorted face of a man in his twenties. There was a big pool of blood behind his head, which rested on a chunk of marble. His face was extremely disfigured and bloodied; his lower lip had been ripped in half, and his jaw was seemingly dislocated. His front teeth appeared to be missing, and it looked as if they had been beaten right out of his mouth. The image was so gruesome that I wondered if he had been wounded in war. But by accessing Dr. Nour's page I learned that Khaled Mohamed Said had apparently been beaten to death on June 6 by two secret police officers in Alexandria.

My first reaction was denial. I could not believe that anyone could actually inflict such brutality on someone else. The victim was a twenty-eight-year-old from Alexandria. According to eyewitnesses, some dispute had erupted between him and the two officers, leading to their physical assault on him, which claimed the young man's life.

I felt miserable, frustrated, and outraged. This was all the result of a political situation that rendered security forces loyal servants of an oppressive regime. Some of our law enforcement personnel had mutated into vicious monsters who were immune from punishment and prone to committing atrocities. They abandoned the Egyptian ethic of goodness that has pervaded our society for centuries.

My memory of that day is vivid. I was sitting in my small study in Dubai, unable to control the tears flowing from my eyes. My wife came in to see what was wrong. When I showed her Khaled Said's picture, she was taken aback and asked me to stop looking at it. She left the room, and I continued to cry over the state of our nation and the widespread tyranny. For me, Khaled Said's image offered a terrible symbol of Egypt's condition.

I could not stand by passively in the face of such grave injustice. I decided to employ all my skills and experience to demand justice for Khaled Said and to help expose his story to vigorous public debate. It was time to lay bare the corrupt practices of the Ministry of Interior, our repressive regime's evil right hand.

The logical first idea was to publish news of Khaled Said's murder on Dr. Mohamed ElBaradei's Facebook page, whose members exceeded 150,000 at that time, but I reasoned that doing so would exploit an event of national concern for political gain. I discovered that a page had been launched under the title "My Name Is Khaled Mohamed Said." I browsed among the posts on that page. It was evident that the contributors were political activists. Their discourse was confrontational, beginning with the page's headline: "Khaled's murder will not go unpunished, you dogs of the regime." From experience I knew that such language would not help in making the cause a mainstream one.

I decided to create another page and to use all my marketing experience in spreading it. Out of the many options I considered for the page's name, *"Kullena Khaled Said"*—"We Are All Khaled

Said" — was the best. It expressed my feelings perfectly. Khaled Said was a young man just like me, and what happened to him could have happened to me. All young Egyptians had long been oppressed, enjoying no rights in our own homeland. The page name was short and catchy, and it expressed the compassion that people immediately felt when they saw Khaled Said's picture. I deliberately concealed my identity, and took on the role of anonymous administrator for the page.

The first thing I posted on the page was direct and blunt. It voiced the outrage and sadness that I felt.

> Today they killed Khaled. If I don't act for his sake, tomorrow they will kill me.
>
> 49 Likes 33 Comments

In two minutes' time three hundred members had joined the page:

> People, we became 300 in two minutes. We want to be 100,000. We must unite against our oppressor
>
> 64 Likes 44 Comments

I wrote the first article on the page: "You People Deprived of Humanity, We Will Extract Justice for Khaled Said." It was an emotional, spontaneous piece of writing. I vowed that I would not personally abandon the fight for Khaled until his attackers were punished. The response was instant, and within a single hour the number of members climbed to three thousand.

> Egyptians, my justice is in your hands.
>
> 50 Likes 39 Comments

I spoke on the page in the first person, posing as Khaled Said. What drove me, more than anything else, was the thought that I could speak for him, and if even a single victim of the regime could have the chance to defend himself, it would be a turning point. Speaking as Khaled gave me a liberty that I did not have on ElBaradei's quasi-official page. It also had greater impact on the page's members. It was as though Khaled Said was speaking from his grave.

Even though I was proficient at classical Arabic *(al-fusHa)* from my school years in Saudi Arabia, I chose to write my posts on *"Kullena Khaled Said"* in the colloquial Egyptian dialect that is closer to the hearts of young Egyptians. For the generation born in the eighties and nineties, classical Arabic is a language read in the newspapers or heard during news reports on television and comes across as quite formal. By using colloquial Egyptian, I aimed to overcome any barriers between supporters of the cause and myself. I also deliberately avoided expressions that were not commonly used by the average Egyptian or that were regularly used by activists, like *nizaam,* the Arabic word for "regime." I was keen to convey to page members the sense that I was one of them, that I was not different in any way. Using the pronoun *I* was critical to establishing the fact that the page was not managed by an organization, political party, or movement of any kind. On the contrary, the writer was an ordinary Egyptian devastated by the brutality inflicted on Khaled Said and motivated to seek justice. This informality contributed to the page's popularity and people's acceptance of its posts.

The number of responses, and the incredible speed with which they came, indicated that administering *"Kullena Khaled Said"* was going to take a lot more time and effort than administering the ElBaradei page. I definitely needed help, and my experience thus far with AbdelRahman Mansour made him the perfect choice. I added him as the page's second admin. During the first few weeks AbdelRahman was quite busy with school and other commitments, but he tried his best to help whenever needed.

I closely monitored news on the case and found the prosecutor's report that acquitted the police force. I wrote:

> The prosecution issued a preliminary report that the cause of death was drug overdose. Not only have you murdered me, but you also want to stain my reputation? God will reveal the truth and repay your lack of conscience.
>
> 55 Likes 112 Comments

Mostafa al-Nagar, ElBaradei's campaign manager at the time, had written a moving article on his personal page entitled "We Are the Murderers of Khaled Said" after he visited Alexandria to verify the story. I published the article on my page without mentioning the writer's name. I did not want people to make the link between al-Nagar and the page and eventually identify the anonymous administrator.

As the page's membership base grew, so did my personal commitment. I felt the stirrings of a rare opportunity to make a difference and to combat oppression and torture. I was angry, and I was not the only one. On its first day, 36,000 people joined the page. Some of them wanted to learn more details about the case, some sought to offer sympathy and support, and others joined out of curiosity because they had received an invitation from a Facebook friend. Images of Khaled before and after the assault spread like wildfire. Similar crimes had taken place in the past, all too frequently, yet their stories had not spread too widely. It was the visual documentation of Khaled's terrible death, along with the fact that he was from the middle class, that catalyzed this huge reaction. The image was impossible to forget, and thanks to social media, it was proliferating like crazy.

By the end of the first day there were more than 1,800 comments on the page. Some people wondered why another page had been launched when the first one, "My Name Is Khaled Mohamed Said," had already reached 70,000 members. "Why not unite our efforts?" they asked. I considered joining forces and closing the page I had created. Yet the aggressive tone adopted by the first page continued to worry me.

I advertised "My Name Is Khaled Mohamed Said" on *"Kullena Khaled Said"* and declared that we all worked for a common cause. I urged people to link to the page and requested that we all coordinate our efforts. To my delight, the admins of the other page reciprocated. It was becoming obvious that this cause could unite a lot of people.

Several prominent opposition politicians publicly condemned Khaled Said's brutal killing. Also, a public funeral for Khaled had been announced for Friday, June 11. I publicized the funeral on the page and asked that as many people as possible attend. I also posted an edited video of various acts of torture by members of the police force, in the hope that Egyptians would finally confront the dark side of the regime and realize that any one of us could be the next victim.

About a thousand people, many of them political activists, took part in the Alexandria funeral. A protest to denounce Khaled Said's murder was also organized in Cairo by the April 6 Youth Movement, among other groups and activists. My hopes for justice were rising steadily. I asked the page members to join the protest, which was planned to take place outside the Ministry of Interior. But the security forces were prepared and decisive: they arrested many protesters and surrounded the rest with double their number of police officers, nearly making a perfect circle. From afar—as later seen in a photograph—the image was quite symbolic. It perfectly represented what the regime was doing to our country. Worse yet, the media, under the usual pressure from State Security, ignored the protest. As with many past examples of human rights abuses, the public was kept in the dark.

The media's suppression of the physical world made the virtual world a critical alternative for promoting the cause. On the Facebook page, I began to focus on the notion that what had happened to Khaled was happening on a daily basis, in different ways, to people we never heard about. Torture is both systematic and methodical at the Ministry of Interior, I said. One of my most significant resources was the "Egyptian Conscience" blog, Misr Digital, by Wael Abbas.

From 2005 to 2008, Wael Abbas actively published every torture document, image, or video that he received from anonymous sources. He was arrested several times by State Security, yet he and other brave bloggers continued to expose the horrifying violations of human rights that were taking place in Egypt.

> I apologize for posting pictures of torture cases, but I swear that I had not seen most of them before. It seems I lived on another planet . . . A planet where I went to work in the morning and watched soccer games and sat at cafés with friends at night . . . And I used to think people who discussed politics had nothing better to do . . . But I am appalled to see a terrifying Egypt that I never knew existed . . . But by God, we will change it!
>
> 658 Likes 188 Comments

I posted links to other torture videos, which were numerous and easy to find. One of them that I published on the page was removed from YouTube, I noticed, because it violated that site's content policies. Many users had reported it as an inappropriate and gruesome video. I did not try to use my employment at Google to resist this decision in any way; my activism had to remain independent of my job.

Meanwhile, although the official press remained utterly silent about Khaled Said's case, the Ministry of Interior began to worry about the controversy. The authorities' first line of defense: stain Khaled's reputation. In an unprecedented public statement, the Ministry of Interior declared that the cause of Khaled's death was not torture but rather asphyxiation, the result of swallowing a pack of marijuana. They said the facial deformation that appeared in the widely circulated photograph was the result of an autopsy. They claimed that Khaled Said had been wanted for four different crimes: drug-dealing, illegal possession of a weapon, sexual harassment, and evasion of military service. As a main player in the state-led defamation campaign, the state-owned *Al-Gomhouriya* newspaper then labeled Khaled Said "the Martyr of Marijuana," a satirical reference to the activists' name for him, "the Martyr of the Emergency Law."

The circumstances of Khaled Said's death were mysterious. According to eyewitnesses, he was sitting at an Internet café when two informers attacked and beat him severely. They then dragged him to the entrance of a nearby building, where they continued to pound him until he died. The official police account alleged that he had tried to hide a pack of marijuana by swallowing it, and that he choked and died while the informers were trying to force him to spit out the pack.

The ministry's expected support of the secret police officers' story, along with the defamation campaign launched against Khaled, exemplified its approach in addressing its problems: never admit guilt, even by a low-level officer. The very limited number of officers who were ever convicted in cases of torture generally returned to work as soon as their prison sentences came to an end.

In response to the ministry's statement, Khaled Said's mother spoke to the independent newspaper *Al-Shorouk* and dropped a bomb: she speculated that her son was murdered for possessing a video showing a local police officer and his secret police colleagues examining and then allegedly dividing confiscated drugs and money. Soon this video, which was allegedly found on Khaled Said's cell phone, spread on Facebook. Many of those who shared it presented it as the reason behind his death. His friends claimed that Khaled had gotten this video by hacking into an informer's cell phone. The video showed a police officer and a few others posing in front of a pile of marijuana and carrying some cash. The officer counted the number of people present and then counted the money and was seemingly about to divide it.

I quickly posted the video, presenting it as a potential explanation for the violence inflicted on Khaled. Yet members responded with disapproval, arguing that my accusations were not supported by clear evidence. I removed the video and posted an apology. It is true that I was quick to accuse the police, and that the officer's actions in the video could have been interpreted differently. The page's members thanked me for seeking the truth and not rushing to defame the police force. Nonetheless, the video spread widely on the Internet and was seen by more than 200,000 users in a few days.

Meanwhile, Khaled Said's family went public with a copy of the military service certificate that proved that he had completed his compulsory service, directly countering an allegation made by the Ministry of Interior. I published the certificate on the Facebook page, as well as videos of three eyewitness accounts of Khaled's murder. One of the witnesses was the Internet café owner, who said that the two secret police officers stormed the place and viciously attacked Khaled. He said he tried to interfere but that only increased their brutality. He also asserted that he did not see Khaled insert anything in his mouth. The second video featured a young boy who saw the beating and testified that others saw it as well but were too afraid to interfere. Finally, the third witness was the porter of the building where Khaled was brutally beaten. He described the viciousness of the violence and said that the officers beat Khaled's head against the stairs while he yelled, "I will die!" But his cries did not deter them in any way. The porter said Khaled lost consciousness and might have died at that point. The ambulance arrived minutes later to carry his body away, without any interference from the residents.

Large numbers of new members were joining "*Kullena Khaled Said*" at unusually fast rates. The page did not belong to any specific patron, and I was careful not to use it for the benefit of any particular political cause, even the seven-demands petition. "*Kullena Khaled Said*" spoke the language of the Internet generation. The tone on the page was always decent and nonconfrontational. The page relied on the ongoing contributions of its members and established itself as the voice of those who despised the deterioration of Egypt, particularly as far as human rights were concerned.

Together, we wanted justice for Khaled Said and we wanted to put an end to torture. And social networking offered us an easy means to meet as the proactive, critical youth that we were. It also enabled us to defy the fears associated with voicing opposition. The virtual world seemed further from the oppressive reach of the regime, and therefore many were encouraged to speak up. The more difficult task remained, though, which was to transfer the struggle from the virtual world to the real one.

I was skeptical about supporting demonstrations, since the first one had had a disappointingly low turnout and had met with such a determined police crackdown. Though many activists had perceived it as a success—since it challenged the might of the ministry—I knew that average young Egyptians, such as the members of the Facebook page, would be easily demoralized if they were treated in a similar manner. Being an activist himself, AbdelRahman Mansour didn't necessarily share that view, but we eventually agreed that it was important not to put our members at any risk whatsoever. So we chose instead to identify online activities that we could promote, to instill a sense of optimism and confidence that we could make a difference, even if only in the virtual world for the time being.

The first campaign I launched suggested that members of the page change their profile pictures to an anonymously designed banner of Khaled Said, featuring him against the backdrop of the Egyptian flag, with the caption "Egypt's Martyr." Thousands responded positively, including personal friends who had no idea that I was the page's founder. Yet some members ridiculed the idea, calling it a helpless tactic in the face of the Ministry of Interior's aggression. The fact remains, however, that our cause gained significant momentum through this awareness campaign.

The strategy for the Facebook page ultimately was to mobilize public support for the cause. This wasn't going to be too different from using the "sales tunnel" approach that I had learned at school. The first phase was to convince people to join the page and read its posts. The second was to convince them to start interacting with the content by "liking" and "commenting" on it. The third was to get them to participate in the page's online campaigns and to contribute to its content themselves. The fourth and final phase would occur when people decided to take the activism onto the street. This was my ultimate aspiration.

I remember debating about all this with Marwa Awad, a correspondent working for Reuters. I, of course, wore my Google hat at the time, and was speaking to her solely as an Internet expert. Marwa believed strongly in the need for change, but like many other Egyptians, she did not think that online activism could create the critical mass needed on the street for achieving real results. People feared the emergency law and the threat it posed to those who opposed the regime or its practices. Yet I was convinced that we could make the leap from the virtual world to the real one. It was going to happen someday, somehow.

The page needed to speak directly to its members and convince them to be active participants, and it was also important to break free from all the barriers of fear that controlled so many of us. So I came up with an idea that served both goals: I asked members to photograph themselves holding up a paper sign that said *"Kullena Khaled Said."* Hundreds of members did so, and we began to publish their pictures on the page. The images created an impact many times stronger than any words posted on the page. Males and females of all backgrounds, aged between fourteen and forty, now personified the movement. The solidarity extended to expatriate Egyptians around the world and to Arabs in many countries—even Algeria, a soccer rival that had defeated Egypt in a World Cup qualifier, leading to heavy violence breaking out among the fans and to feelings of bitterness in citizens of both nations.

Isra, my seven-year-old daughter, had seen some of Khaled Said's photos on my laptop and asked who he was. I explained that he was a good person killed by the police. She innocently said, "Aren't the police supposed to be good? Don't they protect the people?"

"Yes, but some policemen in Egypt are bad," I replied.

Later that day, Isra came to my room to show me a drawing she had made. It showed a policeman shooting at a young man carrying the Egyptian flag. She told me that the young man was Khaled Said.

I hugged her and told her how much I appreciated the fact that she cared about others, and that I was proud of the way she expressed solidarity with them using her own skills. I decided to post the drawing on *"Kullena Khaled Said"* and wrote that our coming generations would not tolerate humiliation and torture.

One picture the page received drove this point home; a pregnant woman sent us an ultrasonographic image of her fetus with a caption that read: "My name is Khaled, and I'm coming to the world in three months. I will never forget Khaled Said and I will demand justice for his case."

The images worked like magic. Members thanked each other for their courage and solidarity. Such admiration and instant positive interaction encouraged even more members to post their pictures. The fact that the regime had not retaliated in any way also made it easier for many people to participate. The barriers of fear were slowly being torn down.

A few days following Khaled Said's murder, opposition newspapers and some private television channels began supporting the cause. I asked page members to apply pressure to TV talk shows. Together, we compiled the telephone numbers of the different talk shows and posted them on the page. I encouraged everyone to call in and demand that show hosts discuss the case of Khaled Said. Earlier, some shows had attacked Khaled, while others had tried to remain neutral. A few had supported his cause, and we were hoping they would now increase in number.

The controversy grew. On June 15, Egypt's public prosecutor transferred the case to special prosecution and ordered a second autopsy to confirm the cause of death. This decision amounted to a small victory for our cause and only served to excite us.

At about that time I noticed an outbreak of comments on the page attacking Khaled, calling him the Martyr of Marijuana, an addict, and a drug dealer. It was such a strong and sudden trend that I decided to investigate. As expected, I discovered that the Electronic Committee of the NDP was behind it. The committee was attempting to convince people that the regime was not responsible for Khaled's death and that he was a dishonorable and unworthy human being. Yet we stayed focused. We were not going to allow such below-the-belt tactics to diminish our enthusiasm and passion for this just cause. It was clear that many page members felt the same way.

> **Mohamed, 26, Alexandria:** How about if we all gather along the Alexandria coast on Friday? We would face the sea with our backs to the street holding hands in silent expression of our disapproval of the injustice inflicted upon Khaled Said. We should try to cover the stretch between the Alexandria Library and Muntazah. It's not a demonstration, but a silent expression of disapproval.
>
> 431 Likes 152 Comments

This idea was sent to the page's e-mail account from Mohamed Eisa, whose full name I did not want to publish on the page so as to avoid endangering his life. I found his suggestion to be very reasonable. A silent demonstration was proactive but not provocative. The general reaction to the idea was positive, and most of the members' comments expressed agreement. I announced the date and time for the following Friday and asked for all suggestions that would help bring the idea to fruition.

Scores of e-mails flowed in to develop the idea. The most important comment was that the effort should not turn into a typical political demonstration, so I called it the Silent Stand, to make the name a clear reminder to all participants that they were not supposed to chant or wave placards or banners. Following a suggestion from one of the members, participants were asked to bring along a copy of the Qur'an or Bible to read in peace. We wanted to send out a clear message that although we were both sad and angry, we were nevertheless nonviolent. Moreover, the way we had planned our stand along Alexandria's coastal road, the corniche—in single file—would have made us inconspicuous among the street's regular crowd, so I passed along another member's suggestion that we all wear black in order to stand out.

The Silent Stand idea was celebrated by many members, who thought it was innovative, yet ridiculed by many others. The activists in particular found the idea too passive and ineffective. They reasoned that if the regime did not care about noisy, dynamic demonstrations, how could it possibly be affected by a silent stand?

It seemed that the majority of page members and political activists shared a common goal but differed on how best to achieve it. Both wanted to send a message of dissatisfaction to the regime. The difference was that the Silent Stand was designed to avoid a physical confrontation with security forces. The goal was for members to summon the courage to take positive action to the street, not just to put pressure on the Ministry of Interior.

The idea is to stand along the corniche, from the Alexandria Library to Montazah, at sunset. We will stand for a half hour and each of us will hold hands with the person on the left and the person on the right, then we will silently pray. For those who are skeptical and asking, "Then what?" this is the strongest message we can send to our government. It says that we reject their practices and at the same time we are peaceful.

363 Likes 82 Comments

An hour after Mohamed Eisa made his suggestion, I created an "event" on Facebook and called it "A Silent Stand of Prayer for the Martyr Khaled Said Along the Alexandria Corniche." The stand was scheduled for two days later, Friday, at 5 P.M.

For this idea to work . . . let's all wear black T-shirts . . . head to the corniche and stand alone . . . turn your back to the street . . . do not debate or argue with anyone . . . this way no one can claim we did anything wrong . . . we want the media to document Egyptian youth standing along 3 or 4 kilometers of the corniche . . . What do you think?

313 Likes 105 Comments

A lot of e-mails requested a parallel stand in Cairo along the Nile corniche. I created another event page for the similar stand at the same time in Cairo. We focused on the safety of the participants, and we stressed the unusualness of the idea, which would make it a strong message to the Ministry of Interior.

> Do you know what is brilliant about this idea? That we are not an organization . . . and we are not a political party . . . we have no motive other than to express our opinion in a civilized manner . . . it is also great because we do not know one another and we are not going to hold a demonstration . . . I swear the whole world will marvel at the Facebook youth.
>
> 242 Likes 40 Comments

I called on Khaled Said's mother to participate in our Silent Stand, to provide a boost of confidence to everyone else. I had no means of reaching her or either of her two remaining children apart from the Facebook page. Egyptians are extremely emotional when it comes to mothers, especially if a mother has lost her son or, worse yet, seen a horrible picture of him dead like the one we all saw.

> A message to Khaled Said's mother: O mother, come on Friday to see your children hand in hand for the sake of your son.
>
> 211 Likes 72 Comments

I was hoping that a member of our page would contact his mother or siblings to convey this message.

Many e-mails from members outside Cairo and Alexandria asked that locations be determined for similar silent stands in other cities. But there was no time to comply, and we had no way to find out how much support there was for the idea in other cities. I wrote this message to the members:

> Guys, I am getting e-mails to plan the event in other cities, but unfortunately we must focus . . . Yet here's a good idea . . . anyone leaving home on Friday should wear black in expression of our silent anger . . . take pictures of yourselves in black and post them on Facebook . . . and—God willing—the solidarity of thousands in Cairo and Alexandria, along the Nile and the Mediterranean, will grace us.
>
> 161 Likes 59 Comments

The individual efforts of the page members were extremely valuable, and they spread further to friends who had not subscribed to the page. The page resembled a product being marketed by its loyal users. For this reason, we relied on the members themselves to promote the event.

> From now until 4 p.m. we must lead the biggest possible publicity campaign . . . We want 1,000 people to help . . . let's spread the links to all our friends . . . anyone who knows journalists must contact them and ask them to publish . . . we will speak to all the TV channels . . . and deliver the invitation to all people.
>
> 1,016 Likes 32 Comments

We wrote a press release to distribute to the media. The purpose, in addition to general publicity, was to mobilize press coverage before the event in order to enable common Egyptians to participate. More than 100,000 members were reached through the page in a few short days. When I compared this with Dr. ElBaradei's page, which reached the 100,000-member mark after more than two months, It was clear that the real cause to fight for was the quest for change. It was change, not individuals, that could unite Egyptians. The merits of any individual could always be scrutinized and questioned, but no one in his right mind could question the need for change. The press release was published on the page, and members were asked not only to distribute it but also to promote the event by designing invitations. A call was put out to all professional graphic designers who were willing to help design logos and banners for the Silent Stand.

The response was phenomenal. And the government was already preparing its response. State-owned newspapers began to attack Facebook by claiming that it was owned by the CIA and that a lot of spies and enemies anonymously used it to brainwash Egyptian youth. The regime despised it, since people were too easily able to express their opposition and discontent. Anti-Mubarak Facebook pages had proliferated after ElBaradei's return to Egypt. I shared this thought with the members to boost their confidence. Our generation was different from its predecessors, who could read only what the regime wanted them to read.

> Do you know why the media are attacking Facebook? Because it does not receive bribes to publish false stories . . . and it does not succumb to security pressure and delete a story . . . Facebook became our means to express our opinions, ambitions, and dreams without pressure from anyone . . . Now our message reaches as far as their biased newspapers . . . But our message is our own . . . We are Egyptian youth who love one another, care for one another, and have a voice.
>
> 383 Likes 85 Comments

The other Facebook page, "My Name Is Khaled Mohamed Said," helped promote the Silent Stand as well. Since its tone was usually bolder and it promoted all kinds of demonstrations and confrontations with the regime, the fact that it published the invitation to the Silent Stand made me quite happy. It meant that political activists would also participate. Activists were more experienced at breaking the fear barrier, and also at handling the security forces in case things got complicated. A young man by the name of Khaled Kamel designed a brilliant video promoting the Silent Stand.

> Thank you to the members of the "My Name Is Khaled Mohamed Said" page for making this video . . . Seriously, the best thing is that we are all cooperating for the sake of Khaled, mercy be with him, even though we don't know one another.
>
> 116 Likes 40 Comments

Sadly, the frustration that prevailed among the Egyptian masses made it necessary to provide boosters of hope as frequently as possible. It was important to make repeated reference to the high responsiveness of our members and to broadcast any and all positive feedback about the Silent Stand. The motto that stuck in my mind as I wrote on the page, trying to transmit its sentiment to everyone else, was "Yes, we can."

> My dear Egypt, your youth are inspiring . . . who can beat 100,000 young men and women who love one another and care for one another? Mohamed e-mailed his idea at 1 a.m. and now there are 2,000 people who plan to wear black and stand silently on Friday. I wonder how he will feel when he wakes up in the morning to discover this. Thank you, Mohamed . . . We are a useful bunch of young Egyptians, I swear, and all we ever needed was a chance.
>
> 323 Likes 132 Comments

The first stand to be initiated by *"Kullena Khaled Said"* was a turning point. I had never exerted so much effort in promoting something, so intensively and in such a short period of time. I worked online for nine straight hours from the time I first posted Mohamed's message, at 1 A.M., until 10 in the morning. No time could be wasted. Our chosen date for the stand was two days later. I was quite stressed, yet my immense passion for the idea served to bolster my energy levels, which in turn were reflected in my posts.

One of the page members visited Khaled Said's mother and invited her to take part in the Silent Stand. She accepted without hesitation, and I immediately posted the good news on the page. As a result, even more people were motivated to participate. I continued to rally supporters until the time of the event. I asked page members to send invitations to the event on Facebook. Invitations were also sent to local and international media to cover the event. The success or failure of the Silent Stand was going to be crucial. If we failed, skeptics would be relentless in their mockery and criticism, which would take us ten steps back.

On Thursday night I had a chat with AbdelRahman Mansour. We were both very excited and optimistic, yet we were afraid of failure. It was hard to predict the number of people who would actually participate, yet it was time to take a risk. We needed another victory, this time on the street and not simply online.

On Friday afternoon, shortly before five o'clock, there was no sign of the Silent Stand at the corniche in Cairo. It became clear that the Ministry of Interior had closely followed the activity on our Facebook page and had prepared itself for this day. Security forces were spread along the corniche. They had closed off some areas and also secured Tahrir Square, in the middle of the city. I found out from Twitter and Facebook that some people did show up in Alexandria, yet the number was very limited. I posted a message on the page, trying to create enthusiasm.

> The numbers are still very limited in both Cairo and Alexandria . . . Where are the people who said they were coming? Where are 10,000 men and women? Hopefully they are only late . . . I hope that none of you are thinking they don't matter because many others are going . . . Please . . . everyone must participate . . . everyone matters.
>
> 121 Likes 113 Comments

I started imagining the level of disappointment that was going to be felt throughout the page. I hate to give up and equally despise the notion that there is no hope, but was there hope, really? This question forced itself into my mind, but I did not dare post it. There was no place for defeatism on a page that had radiated optimism and defiance since its launch.

Then one activist on Twitter said that more than a hundred people were lined up along the corniche in the Cleopatra district in Alexandria. Everyone was wearing black and standing humbly reading the Qur'an or Bible or listening to something on headphones. I quickly updated:

> Guys, one of us in Cleopatra says the scene is different there and more than 100 people have showed up.
>
> 145 Likes 44 Comments

Participants continued to show up in Cleopatra as time went by. My initial worry was not really justified—Egyptians are known to be late, because of traffic congestion.

> People, the scene in Cleopatra has become noticeable, there are more than 300 young men and women dressed in black . . . We're trying to get images from there . . . Where is everyone else? Everyone sitting at home please come out.
>
> 150 Likes 71 Comments

Good news came through Twitter from the Nile corniche:

> There is news that about 20 people are in the Tahrir area . . . But security forces are dispersing any group of more than 3 . . . Guys, let's not turn this into a demonstration, we must all stick to the plan so that today ends peacefully.
>
> 61 Likes 47 Comments

Meanwhile, good news continued to flow in from Alexandria.

Message from Ahmed Fadel: The turnout is huge . . . people were just praying the afternoon prayer . . . there are sooooo many people . . . there is a great increase of participants along the corniche . . . you can barely see the sea from the blackness.

199 Likes 82 Comments

We continued to hear news that the efforts of the Cairo participants were being frustrated in the downtown area, where the security presence had intensified along the corniche. They prevented everyone wearing black from standing along the corniche, which we found to be ridiculous.

Security forces placed barricades and prevented movement around Tahrir . . . Central Security cars and high-ranking officers are keeping everyone from walking . . . are you so scared for your fragile regime that you fear young Egyptians in black?

128 Likes 51 Comments

Shortly after 7 P.M., pictures of the stand were being uploaded by members. Even if the numbers had been small, the impact of a chain of black-clad, silent men and women was impossible to ignore. And in Alexandria the numbers had been large.

AbdelRahman and I exchanged chat messages of joy as we continued to receive excellent snapshots of young men and women standing meditatively in black attire, expressing their anger and their refusal to accept what was happening in Egypt. We were both posting the images on the page. The impact they had on everyone's morale was phenomenal. I also posted a video taken by a page member from Alexandria as she drove by slowly in her car. It lasted a full five minutes, showing one person in black after another. They seemed never-ending, even though there may have been fewer of them than of those who had attended the demonstration for Khaled Said a few days earlier.

We were careful to post every photo we received. Those images helped the participants realize the worth of their achievement, and they drew the envy of nonparticipants. They helped break the fear barrier inside many of us.

We even published an image of four young Egyptians along the Doha corniche in Qatar who participated in the stand. This image was very influential, because it showed that the idea succeeded beyond our borders, not just within them. The extremely innovative idea of these four youths was the reason that many expatriates were later encouraged to join the Facebook page.

Each one of the images we posted carried much greater impact than many days' worth of writing. There is a difference between writing to urge people to do something and showing an image that proves it can be done. Whoever said "a picture is worth a thousand words" certainly knew what he was talking about. Also, such images tend to annoy security forces. Anything that is visually documented is evidence for the whole world to see.

A Reuters news report said that eight thousand people took part in the Silent Stand. I suspect the number was lower, but the manner in which everyone lined up created the greatest impact possible for the number of participants present. It made them appear more numerous than they really were. Marwa Awad, the Reuters correspondent, exchanged e-mails with us and said that the stand gave credence to the possible connection between the virtual world and physical reality. Her support was crucial to obtaining media attention both locally and globally.

Some political activists participated in the stand, and their numbers increased at later stands after the first one proved to be a success. The Facebook page had gained credibility and influence. It had been commonly understood that only a limited number of organizations could mobilize people into action. Many people even thought the Silent Stand was orchestrated by the Muslim Brotherhood.

It is true that movements and organizations like Kefaya, the April 6 Youth Movement, the Muslim Brotherhood, and others were the first to mobilize people on the street. But most of the Silent Stand's participants were young men and women who were not politically active. They had never participated in anything like this before. "The *'Kullena Khaled Said'* page has changed my life": these words arrived in several e-mails from the stand's participants. In general, members' feedback was positive. They felt strong. They had broken the fear barrier. Most important, they had finally transferred virtual activism into real-world action.

Each participant stood silently next to someone he or she probably did not know. They only knew they were both members of a page on the Internet and that they believed in the same cause. They stood a meter apart, wearing black clothes. Feelings of solidarity overwhelmed the participants and turned the stand into a new social environment. We were all concerned for Egypt. We all wanted to abolish torture.

There was a significant backlash nonetheless. Several people were quick to belittle the stand and deem it a waste of time and effort. They asked, "What use was it?" They insisted that nothing would change. Khaled Said died, they said, and no official from the ministry would suffer any punishment. It was a lost cause. For them, the stand was nothing more than a show of foolish, pointless enthusiasm. When such comments increased with every image we were adding, I decided to respond. It was critical to convert as many members as possible from opponents to supporters.

> Many people will think, "So what? What have you gained?" . . . these are the same people who said Egyptians were cowards and no one would show up at the Silent Stand . . . Here are our gains: a strong message that we are a united group of Egyptians who care for one another . . . who are not passive . . . we also delivered a strong message to the Ministry of Interior, the judiciary, and forensics . . . we will expose and scandalize anyone who attempts to torture an innocent person . . . thank you, Facebook youth.
>
> 463 Likes 134 Comments

The contributions from members expressing their real emotions, written in their personal style, worked beautifully. We were careful to add every comment that inspired hope and promoted the importance of the stand; if the stand was not significant to Khaled Said's case, it would at least help transform every member of our page into someone concerned by the events around him or her.

The most important objective was to inspire hope in the hearts of all page members and everyone who participated. "We can" was the critical weapon to fight "There's no hope" and "Nothing will ever change."

Last Friday this page was launched . . . On Tuesday Mohamed sent his suggestion and it was announced to everyone . . . On Friday more than 100,000 members had joined the page and thousands went out in Cairo and Alexandria implementing an idea that was never done before in Egypt . . . So can we do just about anything or what?!

557 Likes 206 Comments

4

Online and on the Streets

I KNEW WHILE WORKING ON the "*Kullena Khaled Said*" Facebook page that I had to be careful to preserve my anonymity. I always had to be the admin, never allowing my true identity or location to slip out. I used a proxy program called Tor which constantly changed my IP address, masking my location by deriving the address from different countries. I also never opened attachments that I received by e-mail unless they were image or text files, which I opened inside Gmail first, without loading them onto my computer. This way I made sure the attachments were virus-free (State Security may not have been full of computer geniuses, but still, I had to be careful). I was also operating on Mac OS, which is, in my opinion, more secure than the widespread Microsoft operating systems. These measures allowed AbdelRahman and me to ignore the threats and insults we received via e-mail, especially since only a handful of people knew our true identities.

On June 19, a day after the first Silent Stand, I decided to develop an opinion poll to find out people's reaction to the stand and to publish the results online. I posted a questionnaire and asked the page members to participate; over five thousand people did. The results helped invigorate the spirit of participatory democracy on the page.

How did security forces react to your participation in the stand?

- 47% They did not even speak to me.
- 32% They spoke with me but I was very calm.
- 15% They spoke with me and I was normal.
- 6% I got angry and they provoked me.

Do you feel the Silent Stand was effective?

- 28% Very effective
- 38% Satisfactory
- 3% Not effective
- 12% Useless
- 19% I don't care

For those who didn't join the Silent Stand, why didn't you participate?

- 6% I was scared.
- 4% I had exams.
- 6% I was too lazy.
- 16% I had work.
- 25% My parents would not let me.
- 20% I'm outside of Egypt.
- 17% I'm not convinced.
- 6% I did not know the scheduled time.

Will you come to the next stand?

- 39% Certainly.
- 39% Most probably.
- 22% I will not come.

Most of the survey respondents were not political activists and were not too familiar with the Ministry of Interior's brutal treatment of its victims. The opinion poll revealed the excitement of those who joined the experience of the Silent Stand and their willingness to repeat it. The poll also confirmed that participants in the stand had in fact been ordinary people, and that many actually did want to take part but had been prevented from doing so by their parents, who feared for them. This was understandable, given the fact that more than 70 percent of the page's members were under twenty-four years old, which meant they were mostly students and still dependent on their families. The weapon of fear that the security apparatus had used for decades had worked with previous generations, but it was becoming clear that it would no longer work with younger Egyptians.

In response, I wrote about the skepticism I had felt, but not revealed, before the stand.

> I want to be honest with you . . . I did not think we would succeed, but I did what I could . . . I even told my wife that I felt I was wasting my time with this stand . . . now I have a story to tell my son and daughter when they grow up . . . and still we have not done anything yet . . . all we did was reclaim our confidence and our certainty that we are united as one.
>
> 593 Likes 118 Comments

Traffic on the page increased remarkably on Friday. Until that day the average number of members who "liked" the daily published content had not gone above 5,000 and the comments had never exceeded 7,000. But on that Friday the number of "likes" reached 37,000, and comments 120,000. This sent me a clear message: action on the ground raised the level of interaction among page members, which was vital for the page's sustainability.

The nature of the comments also changed significantly. The page developed its own culture, and its members began to feel that they belonged to a community. Commenters defended Khaled Said and the page's mission, even though no one knew who was behind the page.

The Ministry of Interior began to understand that its usual methods of oppression might not work in this new world. On June 19, Prime Minister Ahmed Nazif (now on trial for corruption) told *Al-Dustour* newspaper that he was concerned about the Khaled Said case and said that if the secret police officers were convicted, justice would be served. In short, the regime was now trying to co-opt us. It was a partial victory—better than the usual practice of government officials, who systematically avoided discussing cases of human rights violations. This time the intense pressure forced the case into the public sphere. The prime minister's statement about Khaled's case revealed that the pressure applied by private media outlets and on the Internet, in addition to the stands and protests, had certainly captured the regime's attention.

I continued to celebrate every small victory to counterbalance the page's skeptics. Every little step toward justice for Khaled was hailed.

> Why couldn't you behave this way from the start, Nazif? But seriously, I should thank you. If you had punished the culprits of Khaled Said's assault from the beginning, we would not have united and joined forces to fight your monstrous practices against the youth of Egypt . . . From today onward there is no government . . . we are the government.
>
> 456 Likes 200 Comments

Solidarity with Khaled Said transcended national borders. Groups from Tunisia and Yemen began creating Facebook pages in support of our cause. In the face of disputes among Arab youth over trivial matters like soccer, the Internet provided a means to join forces on real issues. Khaled's Tunisian Facebook page attracted over one thousand members within two days of its launch. It was becoming clearer to me that Arabs, no matter how divided they seemed, shared a very deep common anger. Precursors of what later came to be called the Arab Spring were all around us, yet no one knew when winter would end.

This is the new Arab generation . . . A thousand Tunisians came together to support Khaled Said . . . If Facebook youth in this country unite against corruption and injustice, Egypt will become a better place.

470 Likes 68 Comments

The day following the stand I felt the need to create a video that would compile the spectacular images captured by the participants. The video would have to be set to a song that expressed the current situation. Egyptians appreciate and are heavily influenced by art, and I believe that words have more power when they are accompanied by music. After a long search I found "The Resurrection of Egyptians," by Haitham Said. The song was released after the Egypt-Algeria soccer game events, and its lyrics take a tough stance against anyone who thinks of attacking Egyptians. It was intended to unite Egyptians against what had happened, without mentioning Algeria. To me, the lyrics were even more appropriate when applied to the Ministry of Interior.

I had never edited a video before, but I was used to experimentation. After three or four hours of work, the video was ready. I used images from both the Silent Stand and previous protests that corresponded to the lyrics, delivering the clear message that "we, the Egyptian youth, will not give up our rights":

Egypt brought forth strong men . . . Who shined high in heaven
Loyal to their country they have always been . . . Conquering all the fears they faced within.

190 Likes 73 Comments

More than 50,000 members of the page watched the video in the next few days. People found the fusion of images, lyrics, and music inspiring and moving. It was different from the regular practice of lawyers and human rights defenders, who used facts and statistics to garner support. Instead, the video created an emotional bond between the cause and the target audience. Clearly both are needed.

By the end of Saturday, we had decided to schedule another Silent Stand for the following Friday. Several political groups had announced plans for a demonstration outside one of Alexandria's major mosques after the ritual Friday prayer at around 1 P.M. Many of our page members had expressed their excitement about the follow-up stand, especially after seeing the images and the video.

We decided to expand geographically, based on members' requests, so we announced that the next Silent Stand would take place before sunset along the Nile and Mediterranean corniches in ten Egyptian cities. Specific locations were to be determined by the members themselves. I created a new Facebook event outlining a number of demands, and I invited everyone to participate on the designated Friday.

On Friday, June 18, thousands of young men and women in Alexandria, Cairo, and Mansoura left their homes wearing black and expressing sadness, to stand along the corniche with their backs to the street as they read the Qur'an and Bible . . . They prayed for Khaled Said, that God will show him mercy . . . The stand was most successful in Alexandria's Cleopatra and Stanley neighborhoods, and near the Alexandria Library.

This Friday most young Egyptians will be done with exams . . . We want to fill the entire corniche in Alexandria and other cities.

Here are the changes from the last event:

- The time will be 6:30 instead of 5:00 because of the heat.
- Coordination with the media will be much stronger.
- Anyone who will not participate can still show solidarity by sending a photograph of himself or herself standing anywhere for a half hour to mourn the martyr Khaled Said and other victims of torture in Egypt.
- All Facebook initiatives will join forces this time, and God willing, the stand will be a greater success than the last one.

Our demands are

- The immediate judicial investigation of all torture cases
- A public invitation to every Egyptian who has been humiliated or tortured by police officers to appeal securely to the designated members of the judiciary
- Khaled Said's murderers must be prosecuted

But we affirm that

- This is a silent stand. No demonstrations, no chants, no signs.
- This is not a political stand. It is a humanitarian stand in solidarity with victims of torture in prisons . . . As for politicians, we hope they will come, but they are not allowed to take political advantage of the situation.
- We are delivering a message to the entire world that we stand united . . . We love one another . . . We care for one another . . . And Khaled Said is the symbol of what regularly happens to Egyptian youth . . . We will not remain silent again when any one of us is murdered or tortured.

Guys, please spread this invitation to everyone this time . . . like we did last time and even more . . . I hope we all focus our efforts this time to make it a bigger success.

214 Likes 109 Comments

The page members were becoming increasingly eager to know the identity, interests, and ideology of the page's admin. Some were just curious, but others were suspicious. Some comments, but not many, accused the Facebook page of belonging to a political group that wanted to exploit Khaled Said's case for political gains. In order to address this situation, I wrote "Who Are You, Admin?"

Every once in a while I get an e-mail, or someone writes a comment, asking: "Who are you, administrator?" So I decided to answer the questions in a hypothetical conversation.

What's your name?

My name is Khaled Said.
My name is Abdul Samee Saber.
My name is Emad al-Kabeer.
My name is Abdul Razik Abdul Baset.
My name is Ahmed Saber.
My name is that of every Egyptian who has been tortured and humiliated in Egypt.

How old are you?

I cannot say exactly, but since the day I was born our president has been the same . . . corruption has been the same . . . passivity has been the same . . . and a political party called "There is no hope" has been monopolizing Egyptians.

Are you related to Khaled Said?

Khaled Said woke up many things inside of me . . . Thanks to him I now feel I can make a difference . . . Does it make sense that we are not related? . . . He is my brother whom I never saw . . . I wish the informers who tortured him had not killed him so he could see what we are doing for his sake.

Why are you doing this?

When I saw Khaled Said's picture after he was killed I entered my study at home and couldn't stop the tears for some time . . . I saw how fearful Egyptians had become, how tolerant of humiliation . . . I decided to start with myself and change everything I was doing wrong, and this is why I created the page. I cried for Khaled more than I cried over my relatives who passed away . . . And every time I see that picture of him after he was tortured, I get depressed and vow that I will do everything I can to fight injustice.

Who funds you?

Praise be to God, my funding sources are many.

First is my awoken conscience . . . this source allows me to sleep four hours a day and wake up in the morning only to check the page before I wash my face.

My second source is my education. I'm grateful to have been well educated and to have been a top performer academically. I'm using my education to serve my country.

The third source is that I have children and I worry that they might one day ask me, "Dad, why did you do nothing when you saw people get tortured?"

My final source is the love of people who have never seen me, whom I have never seen, and who do not know my name . . . Every day I receive thank-you notes from them and prayers . . . they contribute their designs, logos, and videos, and they promote the page and support Khaled Said's case everywhere.

So what do you want, administrator?

I created this page in secret not only to protect myself from harm . . . but also because I want nothing other than for our country to become ours again . . . I dream that the people will one day love one another again and that none of us will tolerate injustice and that whoever witnesses something wrong in this country will fix it . . . and that an official will think a thousand times before he slaps someone across the face.

Aren't you afraid?

Of course I'm afraid . . . Fear is instinctive in human beings . . . But if there will be people like you who will show solidarity with me as they did with Khaled Said, then my fears will certainly subside . . . we will all meet our fate, and as we say, "No one dies short of life time" . . . A relative of mine died at age 25 in a car accident . . . why should my fate be any better?

Many people will not believe what I have written here, but I swear that it came straight from my heart . . . I really love my country and really wish to see us living under better conditions . . . and I do not wish to start a revolution or a coup . . . and I do not consider myself a political leader of any sort . . . I'm an ordinary Egyptian who cheers the Ahly team, sits at the local café, and eats pumpkin seeds . . . and who becomes miserable when our national team loses a game . . . the bottom line is that I just want to be proud that I'm Egyptian . . . and I don't want any underprivileged citizen to be beaten, not even able to utter his pain.

Forgive me if I've blabbered but I decided to write this so that anytime someone asks "Who are you, admin?" I can just send them this link!

3,761 Likes 2,122 Comments

As with everything else I wrote on the page, I composed the message without planning. I wrote in a language closer to my heart than to my mind. The feedback was great. The message spread widely across the Internet; many of my friends put it on their personal pages, not knowing that I was the writer. I received hundreds of supportive comments and e-mails in reaction to the post. It became the cornerstone of the virtual relationship between members of the page and the anonymous admin.

It was critical to rely on the members to produce content that would engage everyone in feedback and discussion. This was the means to keep Khaled Said's cause alive and to combat torture more broadly. I regularly asked everyone to participate. It was important for everyone to feel responsible and relevant.

I took an oath before God yesterday that I will dedicate at least three hours of my time each day to expose any criminal, thief, or spy in Egypt . . . I will not let this country drown . . . And I really do not want to work alone, I want you all with me, I want you all to feel we must change Egypt.

236 Likes 67 Comments

Calling on everyone granted a talent of writing, drawing, designing, or filmmaking . . . Let us all do everything we can to promote the day of millions in black and the silent stand to mourn the victims of torture in Egypt . . . I hope you can donate some of your time so your work can be seen by thousands . . . We want all Egyptians on the street that day.

221 Likes 70 Comments

When attention is given to talents, they are cultivated further. Members enjoyed it when the content they produced was discussed by thousands on the page. I decided to develop a website dedicated to all the members' poetry, articles, and designs so that nothing would be lost in the future. I called the website "The Martyr": www.elsha heeed.org.

I asked a young man by the name of Amr El-Qazzaz to manage the website. Amr, twenty-one years old, was an active blogger who had once been arrested by State Security for undertaking some independent journalism. He would regularly communicate with me, as the page's anonymous admin, and express his eagerness to help out. His task on the website was to aggregate all the news stories, videos, photos, and even poetry that focused on Khaled Said. The world is certainly small. Amr was destined to play a vital role only a few months later.

A few days after the release of elshaheeed.org, I received a message on the *"Kullena Khaled Said"* e-mail account from Mohamed Ibrahim, an Egyptian living in the United Kingdom (though I did not know his name at the time). He expressed his solidarity with the cause and his interest in helping to spread international awareness of the Khaled Said case.

"Yes, we *need* your help," I wrote, responding quickly to his question. I asked him if we could create an English version of elshaheeed .org and tell the world Khaled Said's story.

Mohamed was very excited and worked for twenty hours straight to set up the English version: www.elshaheeed.co.uk. Shortly after that, he decided to start an English version of *"Kullena Khaled Said"* in Facebook. He wanted to reach the non-Arabic-speaking audience interested in this case and in human rights progress in Egypt.

As soon as the page was launched, I posted its link on *"Kullena Khaled Said,"* praising the anonymous admin of the page and reiterating that this new generation of Egyptians would not stop. Yet I soon received a large number of comments criticizing the page, from Egyptians sensitive about international interference in our country's affairs. It was important to ensure that the majority of our users were comfortable with the idea of an English-language page, so I launched a survey: 78 percent of the more than 1,355 respondents were in favor of communicating our message to the outside world.

Over the next months, the English page played a critical role in helping to gauge support from individuals around the world. Mohamed occasionally posted English letters about the case on the "We Are All Khaled Said" page; this was very encouraging to members on the Arabic page. Mohamed and I communicated and worked together anonymously. I only came to know his real name right after the revolution.

Two days before the second Silent Stand, which was scheduled for June 25, the NDP announced a children's march in Alexandria "coincidentally" scheduled for 5 P.M. along the Alexandria corniche. It was a new tactic by the regime: perhaps they could control the voices of the new youth movement by competing for them rather than crushing them. Traditional police methods had proven ineffective with the first stand—maybe even harmful. They decided to use our own logic and our own tools against us. The difference was that their participants were not self-driven while ours were.

Several political factions, such as Kefaya, the National Association for Change, the Youth of Justice and Liberty, and the April 6 Youth Movement, announced a protest that was to be held outside Commander Ibrahim Mosque in Alexandria just after the ritual Friday prayer. The area surrounding the mosque was packed with thousands of protesters, who were cordoned off by security cars and anti-riot police. The protest was especially successful that day. Dr. Mohamed ElBaradei went from Cairo especially to attend.

I had asked Mostafa al-Nagar if Dr. ElBaradei could join the Silent Stand, since his presence would help put the activities of the "*Kullena Khaled Said*" page in the media spotlight both locally and internationally. After joining the protest, ElBaradei participated in the Silent Stand along the corniche, wearing black like other activists, including Dr. Ayman Nour and the famous anchor Bothaina Kamel. It was a great moment, captured on camera, and even though the Facebook page was never meant to promote ElBaradei as a symbol of change, I posted his photograph standing up with the others in the Silent Stand. It was the first and last time ElBaradei's picture made it onto the page, and it was strictly to promote the cause, with no hidden political agenda.

The second Silent Stand was a huge success. The number of participants increased significantly, and the event took place in ten different areas of Egypt. It was a nightmare for security forces, but for the second time they did not harass any of the participants. Our idea was spreading, and this time it was directly connected to opposition political forces.

We continued to post images contributed by the page's members. One of the participants decided to go into the water and stand for an hour among the waves to express his anger and pain. Another image featured a father, a mother, and a baby that was only a few months old. The father's caption said, "I brought him here so that he would learn to be proactive and not accept injustice." One of the most inspiring photos showed a young man with a broken leg who stood up, holding on to a crutch, for more than thirty minutes. "Despite the pain I suffered, the pain Egypt is undergoing is worth my solidarity," he commented.

I was in the UAE during the second stand. I was supposed to spend the weekend with my children, but I had told my wife that I would have to spend the second weekend in a row working in my study. I spent all of Friday glued to my computer. I followed what was happening very closely, thanks to Twitter and members who used their cell phones to post updates on the Facebook page. I did not sleep at all that night because I was busy posting the images and videos the page received. Hundreds of contributions were coming in, and though we needed to upload them all, we also had to spread the updates out over time to avoid overwhelming the members, since everything we posted reached their personal pages directly.

My wife was upset that I had missed the weekend again. Ever since I had become concerned with politics, I had been spending too much time on social networks. Between my employment and my political activism at home, I was working more than sixteen hours a day. This certainly took its toll on my personal life. But even though Ilka was annoyed, she understood my feelings and knew that I was persistent and would do what I thought was right. And we both saw the second stand increase far beyond the first.

I announced the third Silent Stand for July 9, two weeks after the second one. Part of the reason was to enable the members to catch their breath—and, honestly, I needed to spend time with my family. I even admitted this openly on the page. Many members thought it was wrong to postpone the next stand for a week. They tried to convince me to announce the stand for July 2, but I also wanted time to prepare better for the media coverage. I felt it was important to make the final decision a collective one, though, so we conducted a survey on the page asking members about their preference. A democratic decision was made, and the third Silent Stand was scheduled for July 9.

People who had taken part in the first two stands, and even some who had wanted to but could not, were feeling more and more hopeful. But the pessimistic voices also grew louder. They became my online enemies. Sometimes I gave them more attention than they deserved, but I felt it was necessary—defeatism and loss of hope were like a cancer that spread throughout the whole Egyptian social body. I called these people the "There Is No Hope" political party, and I decided to write a short article called "Abbas and the Administrator" in an attempt to defeat the potential impact of the negative voices.

Abbas and the Administrator

A piece of fiction with a moral!

Please answer this question after reading:

Are you the administrator or Abbas?

I went to see him at the local café after the stand was over . . . It is true he's been my friend for a long time, but we are completely different . . . His name is Abbas . . . A young man in his twenties . . . A worker who minds his own business . . . His most important characteristic is his addiction to passivity and his love for negativity . . . His motto is "I don't care," and his role model is the famous actor who became known for the line

"There is no hope" . . . He believes doing pretty much anything is a waste of time, except to spend all night at the café smoking a water pipe and playing backgammon . . . Abbas is not only a fan of negativity, he also promotes it . . . He is jealous of anyone working for the good of Egypt and criticizes them day and night . . . This way he feels less guilty and believes that since there is no hope it is best to play backgammon!

I found him sitting at the café . . . He seemed delighted . . . I thought that he had finally heard about the Silent Stand and had realized it was a success . . . It turned out that he was happy because Portugal and Brazil tied in their soccer game . . . He likes both teams and wanted them both to qualify for the World Cup . . . I started telling him about the stand, how great it was, how we connected together as young Egyptians who cared for one another.

Abbas gave me a sidelong glance and said, "Hey, admin, what stand are you talking about? . . . You guys need to wake up and smell the coffee . . . Why are you wasting your time on nonsense? . . . It's time to focus on your own life."

Admin: No one said change would come from a stand . . . but at the end, our stand together as men and women, Muslims and Christians, old and young, is proof that we can unite . . . and if our numbers increase, we will surely have real impact . . .

Abbas: Look, you've known me for a long time . . . I swear that no matter what you do, nothing will change . . . If the demonstrations were not effective . . . you think silence will be?

Admin: You know, this stand will be effective! We are not organizing a demonstration that they can ban, and we are not talking or carrying signs that can provoke them to use the emergency law against us . . . and we are not chanting . . . We wear black to mourn the national situation that everyone despises . . . we will wear black until we reclaim our dignity . . . we will wear black until . . .

Abbas: Hold your horses . . . what are these big words you're using? . . . You're making me feel like you're in Guantánamo . . . Our country is just fine, and the fact that something small happened does not mean we have problems . . .

Admin: The problem, Abbas, is that you have become so passive that you have given up on the search for truth . . . I used to think, like you, that there was nothing wrong with our country and that accidents were minimal . . . Then I realized that catastrophes happen on a daily basis, but no one sees them . . . The police torture people systematically . . . the informers threaten the citizens . . . and work with thugs . . . and the law is only enforced on the helpless . . . Anyone who is not well connected in our country loses his rights . . . and when injustice takes place, they threaten anyone who could open his mouth to tell the truth . . . until Khaled Said's case surfaced and we all discovered the truth . . . Look, Abbas, at the videos all over Facebook . . .

Abbas (interrupting again): This is too much for us, and there is no way we can change . . . This is their country . . . don't you realize?

Admin: No, it is not their country . . . it is our country, ours, us Egyptians . . . This is why we must unite so we can change it . . . A nation of 80 million cannot be controlled by a few thousand . . . but we will never exercise our rights if we continue to say "There is no hope" . . . Nothing will happen . . .

Abbas: I want to watch the analysis for the Brazil-Portugal game.

Admin: Believe me, Abbas . . . you know what Egypt's problem is? That there are 80 million Abbases . . . But, God willing, if half a million of them wake up, we will change our country . . . and people like you will sit and watch from the sidelines . . . or if they're unlucky they will suffer Khaled Said's fate before they can even watch . . . but I will not defend you then!

I wonder how many people who are reading this article are Abbases and how many are Admins?

So tell me, which one are you?

688 Likes 735 Comments

The story made members question themselves, and they tried to offer answers in their comments. Later, many of the members quoted the article when they encountered pessimism, saying, "Hey, stop putting us down. Don't be Abbas and start being an admin."

Soon we had a major victory, not just a small one, to celebrate. The regime arrested the two suspected murderers of Khaled Said. The case was overseen closely by a lawyer at the Nadeem Center for Rehabilitation of Victims of Violence, which is an Egyptian civil organization that played a great role in exposing cases of torture and defending human rights for years. On July 1, the Nadeem Center published records of the interrogation of the two suspects. Their testimonies clearly conflicted with each other, and many of their statements were illogical. In response, I wrote an article that argued for the innocence of Khaled Said, based on videotaped eyewitness accounts, together with the informers' statements. It was important to disseminate all of these accounts to try to combat the robust defamation campaign still being waged against Khaled by the official media.

The Silent Stand of July 9 was the first at which tightened security measures were targeted at participants. I guessed that the success of the second stand was behind the heightened police presence, especially because political activists had taken part. Security forces were beginning to realize the true danger of our stand; we may not have been chanting and carrying signs, but we were able to convert silence into strength. I wrote on the page:

> Our silence is strength . . . Our silence is like a whirlwind in the sea.
>
> 329 Likes 67 Comments

A member of the page commented:

> Don't think our silence is weakness, for under the stillness of the earth lies the volcano.
>
> 248 Likes 37 Comments

The designs created for the event featured the sentence "Our silence is strength," turning the abstract notion into a slogan that stuck in the minds of many Egyptians.

A page member in London, Sally Samy, notified me via e-mail that some supporters planned to hold a similar stand in her city to convey their support for Khaled Said. Sally's message made me very happy. Our movement was expanding without becoming centralized, helping us to promote the cause further and faster.

This time around, security forces openly harassed us. Even Alexandria's director of security was on the field in person to inspect the situation that Friday. The police cordoned off the participants and forced them to sit in a circle, surrounded, in a manner reminiscent of typical demonstrations in the past. It was clear that security forces could no longer tolerate us and did not want to give us another chance to make our statement. Many participants in Cairo were asked for their personal IDs, in an attempt to scare them away from future stands, and also perhaps in order to cross-check their records with State Security.

One of the members captured a video of an officer using profane language to order a participant to leave the vicinity immediately. Clips like these incited the page members' anger at the Ministry of Interior even further, particularly because our activism was so peaceful.

Our members were steadily growing in number. Many who saw the public impact of Khaled Said's case decided to emerge from silence and express themselves. They told their stories and distributed documents and visual material in which they condemned the ministry's treatment of citizens. Our volcano was on the verge of eruption, and the ministry could not stop it.

My dedication to the Khaled Said cause, however, came at the expense of my political involvement with Dr. ElBaradei. I was now nowhere near as involved with the ElBaradei Facebook page as I had been. In an attempt to compensate for my increasing virtual absence, I asked AbdelRahman Mansour to give more attention to it. The Muslim Brotherhood decided to officially join the seven-demands petition. The regime was furious, as the Brotherhood could bring a new dimension to the number of signatures collected. Before the

Brotherhood joined the cause, on July 8, more than 100,000 had signed, but once the group joined, the numbers skyrocketed.

State media started to claim that these signatures weren't authentic and to say that since Egypt is a country with 85 million people, if one million people signed, that would still be around one percent of the population. I was closely monitoring the news, and AbdelRahman was making sure to update the ElBaradei page with such news, yet my position of not getting engaged with any political activity while I dedicated time and effort to the Khaled Said case remained the same.

One day in mid-July, I noticed a comment on Khaled Said's page written by a member who was a police officer. I discovered his occupation from the "Education" field on his personal page. He also had friends who were officers, which I could see from their photos, which showed them carrying pistols or posing in their police uniforms. *Why don't we open a dialogue with these people?* I thought. *Let's send them a clear message denouncing the practices of those members of the police force who violate human rights.*

I opened a new document and began pasting the web addresses of the officers' personal Facebook pages. I tried locating the officers by searching for the friends of users whom I already knew were officers and by searching for graduates of the Mubarak Police Academy. After a few hours I had compiled a list of more than four hundred Facebook users who were police officers.

I wrote a post on the page inviting members who wished to take part in a simple yet important campaign to be online within a few hours. It was critical that what I was calling "The Facebook Youth Campaign" be carried out quickly so the greatest number of messages could reach the officers before they had a chance to change their privacy settings. I explained the idea to everyone and warned that we had to remain committed to the ethics of dialogue. Our aim was reform, not collision. Next I planted a fake account in the list of police addresses in order to receive the messages that members addressed to the officers. I also asked the members to create new Facebook accounts to use for this campaign, with different names and a profile image of Khaled Said. The plan was to prevent any personal online quarrels with officers from turning into offline physical risks.

The campaign was successful. Hundreds of the page's members sent thousands of messages to the officers, in addition to requests to add them as friends. My single false account received more than fifty messages and one hundred friend requests! I published the messages on the page for inspiration and as proof that our idea had worked. The dialogue with the officers was unprecedented. Some of them actually supported our efforts. Others opposed us adamantly, referring to Khaled Said as a drug dealer. Ultimately the dialogue helped members further overcome their fear barrier. It also expressed a general commitment to hold the police accountable.

Exactly two weeks after our third Silent Stand was a significant anniversary on the Egyptian calendar. July 23 was the date of the 1952 revolution. We were in luck that it fell on a Friday. With ten Egyptian cities participating, the third Silent Stand had succeeded in broadening the movement, but the turnout at each location had not reached the level of the previous stand. Now, with the anniversary and the upcoming trial of the two suspects accused of murdering Khaled Said (set for July 28), Friday, July 23, offered a chance both to expand participation and to direct public attention to the trial.

Some historians believe that the 1952 revolution and the ongoing military rule it ushered in were the reason behind Egypt's deterioration. Nevertheless, the anniversary of the revolution was a national holiday, featuring formal celebrations and a speech by President Mubarak. The media typically covered these events all day. I announced on Facebook that our fourth Silent Stand would take place on July 23. I now called it "The Revolution of Silence."

Many commenters mocked the name, thinking of it in relation to our forefathers' revolution against the monarchy. Their jokes filled the page, declaring that the idea was useless at best and maybe even harmful. But we decided to go ahead with the stand and rally for support.

The notion of linking the Silent Stand to the July 23 revolution was particularly disturbing to the security apparatus. The government did not wish to be embarrassed in any way. I began receiving threats by e-mail. One message came from a girl who claimed her father was a State Security officer and that they had discovered my identity through my IP address. I had to cease my activity, she said; she was worried that something might happen to me. Yet I knew they could not possibly discover my IP address, because I was particularly careful to use the proxy.

In 2006 I had seen the movie *V for Vendetta* and fallen in love with the idea of the mysterious warrior fighting against evil. I was still influenced by this idea when I created the Facebook page: the notion of an anonymous sentinel who tries to wake up the people around him and spur them to revolt against the government's injustice. For my article "Who Are You, Mr. Admin?" I used the distinctive mask worn by the movie's protagonist as the main image. I even posted clips from the movie to promote the Revolution of Silence. I identified with V's desire for change, although in no way did I approve of his violent means. Yet the image was quite symbolic of the admin's anonymity.

I downloaded the movie from the Internet and cut out an important scene, one where V breaks into a television station and cuts through the propaganda to appear live, on-air, and address the people, exposing the reality of their situation. I translated the speech in which he tells his fellow citizens to take to the streets on November 5 in revolt against injustice. I remember this line: "If you're looking for the guilty, you need only look into a mirror." I loosely paraphrased it and appropriated it to invite Egyptians to come out in silence on July 23 in a revolution against the Ministry of Interior's corruption.

I posted many of the usual guidelines in preparation for the stand, and some members contributed new designs and video clips promoting it. Khaled Kamel, the twenty-one-year-old who had edited a promotional video for the first Silent Stand, made an exceptional effort to rally supporters before the July 23 stand. I frequently corresponded with him anonymously about new creative contributions, and I was cautious not to mention any details of my life during our chats. Yet something very worrisome once happened in the middle of one of our conversations.

We were chatting about a video I wanted him to make, and he sent me a link to his blog. I clicked through because it was hosted on Blogger, which belonged to Google. But then he surprised me by telling me that he had found out where I lived. His blog used a plug-in to identify IP addresses, and I had, unfortunately, forgotten to use the proxy. "You live in Dubai," he told me. "So this is why you feel safe. No one can catch you."

I felt myself turning red and beginning to perspire as I read what he wrote. The problem was that with my IP address in hand, he could easily locate my home as well. I tried to defuse the situation. I sent him a smiley face and told him I was using the software Tor: I could have been in Norway, or Iceland, or Switzerland, and the software would hide me. He did not believe me and asked me to relax, as he didn't plan on telling anyone. I turned on Tor and accessed his page again, showing him an IP address in Latin America. I did it again and he got an address from Japan. At this point he appeared a little more convinced, although not fully. The last thing I wanted was for someone I did not know to obtain details about my personal life. But to my relief, this episode ended in peace.

We heavily promoted the July 23 Revolution of Silence in many different districts of Egypt. So did other political movements and Facebook pages, such as "My Name Is Khaled Mohamed Said." The turnout was lower than expected, but even so, something happened that day that had not happened before. Some political activists in Alexandria decided to march to Khaled Said's home to chant for his mother, vowing that justice would be served and that we would all stand with Khaled Said. I did not support the march because I thought it was too risky for the participants. I felt responsible for demonstrators who were not political activists. They were my brothers and sisters, and

I did not want them to be harmed. AbdelRahman Mansour, on the other hand, saw that it was important for them to take ownership of what was slowly becoming a national project for change.

The march reached Khaled Said's home, and the narrow side street was packed with protesters. The chants began with "Kullena Khaled Said" and cries for justice. Then the chants turned into calls to attack the interior minister, Habib el-Adly: "If Khaled was a minister's son, el-Adly would be executed." The tone quickly hardened, and the common antiregime chant began: "Down, down with Hosni Mubarak." The videos captured by page members showed that this chant was louder than the others. I was a little angry at the activists. It was not as if I liked Hosni Mubarak, but I did not want to lose the support of those page members who would find a direct attack on the president to be either inappropriate or very risky or to create a conflict between the page members who held different opinions; the page had to do its best to stay focused and gather people around the cause.

Clashes took place at the march, and some security officers turned violent in order to scare the demonstrators and crush the event. This left a few of the participants feeling angry and emotional, so I wrote several posts to show that the marching and chanting were the cause of these confrontations and that it was better to discipline ourselves so the Silent Stands could proceed peacefully and attract more participants in the future. I also wrote an article under the title "Down, Down with Hosni Mubarak," in which I tried to explain that even though I did not support the president, I believed that chanting against him was not wise. The Khaled Said case should focus on the rule of law and human rights. Making it political could cause us to lose public support. In response, some members attacked me, calling me intellectually immature. They said I lacked political vision, that I didn't realize our political situation was the cause of all our problems.

An activist from Alexandria wrote a critical piece declaring that the people, and not someone sitting behind a screen, should make these decisions. He said that the nonpoliticized participants were the ones who began chanting against Mubarak and that I must let the people decide for themselves how to act. I published his opinion despite his criticism because I wanted him to have the chance to convince others. A noticeable percentage of the members seemed to agree with his views. I believed that the risks we were taking by pressuring the Ministry of Interior were proportional to the potential gains we could achieve. Becoming confrontational with the regime would create too much danger for too small a reward. At the time, I did not know whether my opinion was right or wrong, but at least my conscience felt at ease. I did not want to expose the people I mobilized to any security threats.

Some of the Silent Stand participants in Cairo were harassed that same day. A police officer had hired thugs to disperse the participants at the Nile corniche, and one young woman captured a thug on video as he cut off tree branches with which to beat and scare demonstrators. He saw the girl, rushed toward her, and batted her phone away, sending it crashing to the ground. The image went black. We only heard the girl screaming and yelling at him.

The risks of the Silent Stands were becoming more evident. The participants were growing eager for confrontation—they wanted to protest and chant and not just stand silently. In an attempt to calm everyone down, I posted an article trying to explain why the police attacked the protesters.

Why Do They Attack Us?

It is important to place ourselves in the position of our adversaries and try to figure out why they behave the way they do . . . This will also help us determine the appropriate response and come up with ways to influence their decision-making.

Let us first discuss the people who order the attack, not the ones who carry it out.

First, in order to fight an idea you must have an opposing idea . . . The fast way to fight any idea, from an oppressive ruler's point of view, is to scare the people so they don't think . . . This method is low-cost because it relieves them of the burden of accountability. Beating people is a shortcut . . . A scared citizenry will not speak or object . . .

The objective is not to scare people who attended the demonstration in order to keep them from participating again; police officers know well that demonstrators who were attacked once will come back again and again. The objective is to scare the people at home so they never engage in political activity in the first place. The security apparatus is applying the old Egyptian saying, "Beat the tied one and the free one will get scared." The government is always afraid of the snowball effect. If today 10 demonstrators show up, then maybe 100 will show the day after and 1,000 the day after that, and so on.

For the regime, one of the greatest advantages of attacking and beating protesters is that it increases their animosity. They want protesters to lose their temper and reject any defense of the regime . . . They want you to curse the government in the most profane language . . . Why? Because then the media can use you as evidence that we are all irrational extremists . . . This way regular people who steer away from politics will steer away from you as well.

Ordinary Egyptians are convinced that most people who go out to protest against our country's condition are really aiming to create chaos . . . They think that protesters are controlled by spies and are merely posing as patriotic. Most of them take this for granted and refuse to reconsider. In their minds, it is as solid as the fact that the October War was in 1973.

Ask one of them, "Is there or is there not corruption in Egypt?" He will say, "Yes, of course, there is a lot of corruption, but still we must defend the stability of our country. Without it, there would be chaos, and chaos could lead to even greater corruption and mayhem. So we must not give these dissidents the chance to express themselves."

Officials at the Ministry of Interior think that all protesters are funded by the Americans because the U.S. government wants to readily be able to pressure our government whenever needed. And, unfortunately, ministry officials have only become more convinced after the American secretary of state's comments on Khaled Said and the arrival of the American attorney general and members of Congress, who came to Egypt to demand that the government lift the emergency law . . .

What's the point of all of this?

We must understand how these people think so we can

always stay a step ahead . . . We must abandon the notion that some people are absolutely good and others absolutely evil . . . Every one of us has faults and some of us even have the same faults as the government, but the difference is that they have authority and we don't . . .

We must diversify our methods . . . Because to continue using one method—namely, protest—will not lead us to the change we seek.

1,111 Likes 272 Comments

Inspired by Gandhi and other advocates of nonviolent resistance, I was keen to stress regularly that our activities were to remain peaceful at all costs. Gandhi is certainly one of my heroes; I have enjoyed reading many books about his philosophy and about how he revived ahimsa, the ancient Indian religious principle of nonviolence toward all living things.

Quotes from Gandhi frequently came up on *"Kullena Khaled Said."* I even once called upon our members to watch the biographical feature film *Gandhi,* produced in 1982. To me, one of the most inspiring scenes in that movie is when Gandhi addresses Indians in South Africa about the next move in their struggle for basic human rights. At one point, he invites everyone to strike against some newly passed and heavily discriminatory laws, and some of the people passionately affirm that they are prepared to die for the cause. Gandhi replies, "In this cause, I too am prepared to die, but my friends, there is no cause for which I am prepared to kill. Whatever they do to us, we will attack no one, kill no one, but we will not give our fingerprints, not one of them. They will imprison us. They will fine us. They will seize our possessions. But they cannot take away our self-respect if we do not give it to them." A member of the audience shouts: "Have you been to prison? . . . They beat us, they torture us." Gandhi interrupts: "I am asking you to fight. To fight against their anger, not to provoke it. We will not strike a blow, but we will receive them. And through our pain we will make them see their injustice. And it will hurt, as all fighting hurts. But we cannot lose. We cannot. They may torture my body, break my bones, even kill me. Then they will have my dead body—not my obedience!"

I translated this entire scene and posted it on the page. Gandhi's triumph in the face of the British Empire assured me that great battles could be fought and won without violence.

Adopting a nonviolent and nonconfrontational approach was not the only theme that we consistently declared on "*Kullena Khaled Said.*" The page also relied on participatory democracy in making most of its decisions, particularly those that involved activities on the ground. I have always been a firm believer in engaging and empowering audience members, as that will increase not only their confidence but also their desire to promote the cause. Engagement was the page's core concept and was certainly far more important to the page than activism.

On July 25, we launched a survey to learn what active members thought of the page and how it was managed. The survey was posted several times, and within five days more than four thousand users had completed it. As expected, over 81 percent of the respondents were less than thirty years old. More than half of them were aged between eighteen and twenty-four.

The key findings confirmed that the page's strategy sat very well with the active members. Seventy-two percent of the respondents said that they strongly admired and respected the page, while 75 percent said that they felt like they owned the page and that the causes it promoted were their own causes. Only 14 percent believed that the admin was a dictator who forced his point of view on members, and 7 percent thought the admin contradicted himself often. This was strong confirmation that we were on the right track.

Five days after the July 23 Silent Stand, it was time for the first court session of the Khaled Said case. The two informers were being accused not of murder but of unjustified violence. The prosecutor still claimed that the cause of death was asphyxiation resulting from swallowing a package of marijuana. Activists decided to hold a symbolic stand outside the court before the session. Our Facebook page endorsed this and publicized the event.

When the activists arrived outside the court, they found scores of people who had clearly been mobilized by the regime. They were carrying placards, all designed by the same calligraphy artist, with perfectly coordinated messages: "The police serve the people"; "Khaled Said is the marijuana martyr"; "No to foreign agents – wake up, Egyptians." They were trying to convey that Khaled Said and his family were agents of foreign powers, as though his murder would have been justified if that were the case. As the number of activists increased, some verbal battles erupted. When the court session began, anger was still humming in the air.

The first court session was strictly procedural. Each side requested witnesses; Khaled Said's family rejected the forensic report and requested that the prosecution charge the two suspects with murder. The judge decided to postpone the case until September to summon the witnesses and examine the documents. A wave of disappointment was expressed through the Facebook page, especially since most people, including AbdelRahman and I, knew little about judicial procedures. We tried to raise everyone's spirits by highlighting the fact that the case would have been closed if not for the popular attention it had received. We stressed that even seeing the suspects brought to trial was an accomplishment.

But the trial was postponed for procedural reasons at every subsequent court session. Modern Egyptian judicial procedures are extremely slow and bureaucratic. Every postponement pushed the trial date back another month: from September to October, to November, to December. Everyone started to lose hope.

Around the same time, Amnesty International (in coordination with Khaled Said's family) released a consultative report written by international forensic experts. The report detailed the Egyptian forensics team's errors and failure to follow international autopsy standards, casting doubt on its initial report. Amnesty International's

Egyptian employees were following the Khaled Said case very closely, and the organization came out in support of bringing the case to trial.

Three hours each day were no longer enough for me to keep up with the Facebook page. It was like a hurricane that engulfed most of my time. I became less focused at work, less willing to discuss, debate, and fix every problem I encountered. I was putting in just enough effort to get by, but the quality of my work had deteriorated dramatically. Fortunately, Google's management was not alarmed by my drop in performance: they thought that I was finally balancing work and family. Little did they know that I had given up all family time and had started using work time for my Facebook activism.

My wife was used to the fact that the weekend offered her only chance to really communicate with me. But because of the Friday Silent Stands and the intense preparation efforts that preceded them, I had to make up for lost sleep on Saturday. Ilka complained a lot about my preoccupation with the Facebook page and the political activism that was taking its toll on my family life. She said my family was not at the top of my list of priorities, and she was right. I promised her that I would spend less time on *"Kullena Khaled Said,"* and I asked AbdelRahman Mansour to help me out more by taking care of the page on Saturdays.

I remember one particular Saturday when I took my wife and children to the movies, as I had promised to do following the Silent Stand on Friday. While they watched the movie, I was busy reading and posting on the page, using my cell phone. When the movie ended, it was obvious that Ilka was upset. When I asked her why, she answered that my attention had been somewhere else the whole time. I had not watched a single scene from the movie, she said.

I tried to defend myself. Wasn't I helping to bring about change in my country? But I couldn't convince her. The truth is, I couldn't even convince myself. Putting myself in her shoes, I knew that she was right, but I was having a personal struggle because Egypt was experiencing such a critical time. I was trying to do my part, to repay the debt I felt I owed to Egypt by trying to bring the case of Khaled Said into the public consciousness.

The "My Name Is Khaled Mohamed Said" page, the one that used a more belligerent tone and appealed to activists, began to grow stronger over time. It had 60,000 more members than we did, and its aggressive tone created problems for us, because some users did not distinguish between the two pages. Right before one of the Silent Stands, the other page's administrators decided to convert the event into a demonstration. So I decided to contact them for the first time and discuss their approach. It was very easy to discover their names, even though they weren't publicized on the page. After my first e-mail, I was replied to by someone who called himself Mahmoud Samy.

I searched for him on Facebook and discovered that he was an April 6 Youth Movement activist and a university student. We had many friends in common, mainly activists. I sent him a message, and we agreed to have a voice chat over Gmail. I requested that he not ask any personal questions or try to determine my identity. He agreed.

At the beginning of our conversation I asked him for some of his page's statistics. When he replied, I ran a quick comparison between his stats and ours. Even though our page had 30 percent fewer members, the participation and comments on our page were sometimes double the other page's. I asked him if he knew why that was. He said he did not know but added that he had noticed that users liked "*Kullena Khaled Said*" and were usually positive and responsive toward the admin's posts. I found his response delightful, as it gave me the perfect starting point for what I wanted to discuss.

"The issue lies in the difference between activists and regular non-politicized young men and women," I began. I went on to say that activists speak in rebellious language that is hard for those who have not gone through similar experiences to understand. The result is a gap between activists and their audience. Activists' ability to mobilize and persuade becomes limited to a specific segment of people.

"Let me give you some examples," I said. "The expression 'dogs of the regime' has been a central feature of your discourse since the page was launched. I agree that this is an appropriate, angry description in light of Khaled Said's horrible fate, but there are certainly groups of members who find this kind of language repulsive." The tone of the page was always angry and rebellious. I asked Mahmoud if he studied marketing, since he had told me that he was a business undergraduate. He said no. So I went on to explain that every brand has its own characteristics. Both Facebook pages were brands. Our page, "*Kullena Khaled Said*"—our brand—was peaceful and inclusive; it sought justice and involved its participants in decision-making. His brand, "My Name Is Khaled Mohamed Said," was rebellious, angry, sometimes ill-mannered, and often dogmatic. He agreed.

"Why do you use the plural *we* instead of *I*?" I asked.

"Because *we* is more professional and credible," he said.

I told him about my experience with the first-person singular. It has a stronger impact, I said. The plural *we* makes readers think an organization is behind the page, devising plans and putting them in practice. *I* simplifies matters and opens the way for direct dialogue between the members and the page's anonymous admin. I also mentioned the benefits of using colloquial Egyptian, and explained the importance of apologizing when mistakes were made and of presenting opinions credibly and objectively. In addition, I highlighted the value of thanking police officers when they did something positive.

Our conversation lasted over an hour. It was an intense marketing crash course, and the good news was that Mahmoud Samy appreciated it. Ever since, he has jokingly referred to me as his "Internet godfather."

Afterward, I began to see some changes in the way "My Name Is Khaled Mohamed Said" was run, but the shift was not comprehensive. Mahmoud Samy was one of four administrators who managed the page, each with a different style, and he was able to convince only one other administrator to adopt the approach I had suggested. I remained in touch with him and continued to offer guidance, but he could not overhaul the page by himself. At one point he even wanted to terminate his affiliation with the page, but he chose not to for the sake of maintaining even the partial influence he had on its brand image.

Then, on September 8, 2010, Facebook's management decided, without prior warning, to terminate the administrators' ability to update "My Name Is Khaled Mohamed Said." When Mahmoud and the other administrators asked for some justification for this decision, they were told that the page had violated Facebook's terms of usage by uploading copyrighted material and that the decision to block its admins was irreversible. This served to increase the responsibility of our page.

During September, our page reached the 250,000-member mark, after less than three months of operation, even though we did not advertise or make any media appearances. We celebrated this milestone. The Facebook page had become a resource powerful enough to compete with official media for readership. "*Kullena Khaled Said*" members, however, tended to be young, while the older citizens who influenced political events in Egypt remained far from our page. This gap, or "digital divide," turned out to be useful later, when people started promoting the revolution. No officials were able to predict the size of the turnout.

We were still a long way from that, though. In September 2010, the Khaled Said case was on hold and the Silent Stands had lost their appeal. We needed to find some other way to keep the members engaged and united and to continue pressing for change in Egypt. It was important to make them feel proud of being proactive. We also needed to promote the culture of dialogue and acceptance of the "other" and to cultivate a tradition of tolerance among page members.

My own enthusiasm was beginning to fade. It seemed like I had done all I could do. I no longer wanted constantly to write posts on the page. I was always careful not to write something that I did not genuinely feel. If I felt upset and pessimistic, I did not try to appear optimistic and happy. I believe that messages from the heart reach the heart and messages that are forced appear for what they are: artificial. As a result, the participation level on the page dropped between

September and November. There were hardly any campaigns or initiatives other than following up on news, particularly about Khaled Said's case.

I didn't know what to do. I discussed various ideas with Abdel Rahman and some other activists. One idea was to organize a silent stand expressing our disapproval of the coming parliamentary elections in November. There were many clear signs that indicated the regime's intention to rig the elections. But everyone I spoke to rejected the idea — it would not produce any results.

Then Khaled Kamel, the young man who produced videos for our page, sent me clips of protests in Chile that had extended for months and ultimately succeeded in overthrowing the ruling regime. Some of these protests were based on the idea of using any tool or device available to create noise and annoy the people around you: honking car horns, banging on cooking pots, running kitchen appliances, clapping, whistling — anything. The point was for a large group of people to take to the streets, attract attention to a certain issue, and break the fear barrier.

We posted the idea on the page, but reactions were mixed. Khaled Kamel put together an excellent video explaining the idea, but most commenters said, "We're not like Chile — we have a different culture." Some members said that noise was not a civilized means of expression. The activists, however, liked the idea and thought it was a step up from the Silent Stands.

I created an event on the page and sent out an invitation. We did not specify any location, though some activists, mainly from the ElBaradei campaign led by Mostafa al-Nagar, chose to protest in Imbaba, a populous but underdeveloped neighborhood in northern Cairo, as well as in one of the major neighborhoods in Alexandria.

But when I was creating the event, I made a terrible mistake. About half an hour after I set it up, I realized I had misspelled a word in the event's name. By the time I had corrected it, more than five hundred people had already confirmed attendance. The typo was not my terrible mistake, though. I soon received an e-mail message from

Mahmoud Samy, the admin of "My Name Is Khaled Mohamed Said," telling me that he had discovered my identity. He included an attachment of a Facebook notification he received that said, "Wael Ghonim has changed the event name." Without realizing it, I had been using my personal account when I created the event, not my anonymous admin account.

It was a horrifying moment. Mahmoud told me not to worry, and he swore he would not tell anyone. He implored me to hide my name somehow. I started to sweat. My heart beat faster. The five hundred RSVPs might include some from State Security officers or even people who knew me personally. In either case, my identity might be leaked. I tried to calm down and think rationally. I finally realized that I could change my Facebook name to Mohamed Ahmed, so users would see that pseudonym instead of Wael Ghonim. But even when I did that, the link still took users to my main page, where the username was ghonim. So I decided to disable my Facebook account temporarily. At that point, the alert had changed to "Mohamed Ahmed has changed the event," and it no longer linked to my Facebook page. Yet I remained petrified.

I spent a terrible night thinking about what could happen if someone discovered my true identity. I told my followers on Twitter that I was unable to access my Facebook account because of a technical problem, so in case I was interrogated I could use this as proof against my involvement with the page. I would claim that my account had been hacked by someone trying to ruin my reputation. Still, I did not sleep for a minute that night. I chatted with AbdelRahman Mansour to explain what had happened and seek his advice, and he shared my concerns. The biggest problem was that a well-known activist, Mahmoud Samy, now knew me by name. I created a fake Facebook account and removed the admin privileges from my real one. I then wrote to Mahmoud Samy asking him not to reveal my name, partly because my life would be in jeopardy, but more because I didn't want people to put a face to the person who had created the page. The movement needed to remain focused on Khaled Said and the public as a whole, not on me. He swore that he would not tell anyone and that he would forget the name and all the details associated with it. I now felt even more respect for him.

Yet fate had more anxiety in store for us. Two days later AbdelRahman committed the same error. He wanted to edit the event details again, and even though I had told him how my slip-up had happened, he forgot and did the same thing! I received an alert with *his* name as editor of the event, and I called him, raging. What I did was the result of negligence, but what he did was much worse, I said. How could we make the same mistake twice in two days? It was as though we were handing ourselves to State Security on a silver platter.

One of the activists on the page, whose nickname was "How Dull," sent a message to the page's e-mail account saying, "AbdelRahman, your name was revealed. Please do something." I tried to calm her down and said that my name was not AbdelRahman and that I used a pseudonym. I asked her not to worry, but she still seemed quite concerned.

Next I took away all the admin privileges from AbdelRahman's account and asked him to create a fake account and close down his real one for a week. This way his real name would not appear in the Facebook alerts sent to people who RSVPed for the event. Nevertheless, these few short days were a nightmare. I was obsessively worried that State Security had already found out who we were.

Turnout at the Day of Anger protests was much lower than I had anticipated. Many people found the idea unappealing, particularly since the protest locations were not announced in advance and thus the critical mass required for real effect could never be reached.

As the parliamentary elections approached, it didn't take AbdelRahman too long to convince me that the page should cover them. We decided to stay as neutral as possible. AbdelRahman also came up with another brilliant idea, which was to call on any page member who was a registered voter to nullify his or her vote by voting for Khaled Said. This was a proactive way to boycott the elections, and it would help reduce the number of forged votes. I really wished we had time for a full-fledged marketing campaign that would help spread the idea widely, but by then the elections were only forty-eight hours away. Nevertheless, better late than never. A volunteer graphic designer created a profile page with a call to vote for "Khaled Said, the symbol of Egyptian youth."

We planned on exposing any violations regardless of who committed them. The NDP had been publicizing its expected sweeping victory, and as usual the state-owned media machine was viciously attacking the Muslim Brotherhood. All the relatively large opposition political parties and groups announced that they would boycott the elections, with the exception of the Muslim Brotherhood and Al-Wafd Party. A rumor had spread that both had cut a deal with the regime in exchange for a limited number of seats in Parliament, but it proved to be false.

Then, suddenly, one day before the elections, the *"Kullena Khaled Said"* page was suspended by Facebook. I tried accessing the page but was redirected to the main Facebook page, and I had no idea why. I noticed that the same thing had happened to a few other pages as well. The news spread quickly on Twitter; everyone was tweeting that our page was closed. Many Twitter users sent messages to me as Wael Ghonim, asking me to use my Google connections to reach Facebook's management and retrieve the page for the administrator.

I received a message from Nadine Wahab, a U.S.-based Egyptian human rights activist whom I had met online through e-mails when we first launched Dr. ElBaradei's campaign. She was eager to help Egypt's quest for change in any way. She told me that she wanted to help retrieve the page and that she was trying to get in touch with the management at Facebook. Nadine was one of the very few people who knew my real identity. A few months before this incident I had deliberately told a few activist friends who lived abroad that I was the admin of the page so that they would take the necessary actions if my life was ever at risk.

AbdelRahman was becoming very frustrated and anxious as well, trying to find a way to get the page back. "If only I hadn't changed my account! I can't believe we lost the page and all our long months' worth of work," he said. I was much calmer than he was, because I was confident that the page would return. It would be best for both of us to calm down, I said.

Finally Facebook replied to my inquiry. The admins said that the page had been closed after a lot of abuses were reported and it was discovered that "*Kullena Khaled Said*" was managed by fake accounts, which violated Facebook's policies. Our attempts to remain safe and anonymous had come back to handcuff us.

Meanwhile, Nadine Wahab's efforts led her to one of Facebook's senior officials in Europe, who was responsible for the company's policies in Europe and the Middle East. The agreement they reached was that the page could be retrieved on the condition that the page owner must be a real person's account, without aliases. After long negotiations in which I was involved as the anonymous admin, we reached a deal: Nadine Wahab and AbdelRahman Mansour would become the official page administrators. Although Facebook was not going to make any admin identities known to the public, I opted not to include my own. I wanted to avoid any sensitivities that could arise as a result of my being employed by Google.

Nadine showed great courage and passion in offering her personal account to me to use to manage the page. She said she lived in the United States and had no plans to return to Egypt anytime soon. She was also willing to take the risk if any problems occurred. I asked AbdelRahman to create a fake account so that I could grant him access once again. I also created a fake account to make sure Nadine's account was not suspended in case this problem occurred a second time.

Remarkably, even though "*Kullena Khaled Said*" had been closed for less than twelve hours, the news had made its way into many Egyptian and a few international headlines. People connected the page's suspension to the elections. Conspiracy theories flourished, suggesting that Facebook had struck a deal with the Egyptian government to block certain activist pages and provide detailed information about the people running them. These accusations were later proven to be entirely baseless.

We were relieved when the page was live again. And it had recovered its value in our eyes: it was a page of 300,000 users, most of whom were young Egyptians who believed in the dire need for change and reform in Egypt. What a blessing!

The elections began, and so did our monitoring of violations. We asked our members to report any transgressions they witnessed. Election fraud was the default practice in Egypt. There were usually thugs outside the polls, who would terrorize voters before they entered to submit their votes. Muslim Brotherhood supporters were regularly beaten. The names of deceased citizens were used to cast false votes, and on occasion people who hadn't voted found that their identities had been illegally appropriated and used in the ballot boxes (this happened to a friend of mine). Finally, State Security would pressure the civic workers responsible for sorting and counting the votes into rigging the results.

Even so, the 2010 parliamentary elections were the worst in Egypt's history. A group of young people sympathetic to the Muslim Brotherhood launched a Facebook page called "Monitoring—2010 Parliament." It was delightful to see a page devoted to citizen journalism appearing on Facebook specifically to monitor the elections. They called themselves Rasd, an Arabic word that means "monitoring." Judging from the way the page was run, its founders seemed both professional and young. I would later find out that one of the founders was actually none other than Amr El-Qazzaz, the guy to whom I had delegated the management of elshaheeed.org. What a small world! What's more, fate had an even more critical collaboration in store for us only a few weeks later.

The "Monitoring—2010 Parliament" page attracted more than 40,000 people before the elections even started. The administrators developed a network of correspondents in all electoral districts to report violations that occurred. Our page did the same, but in a less focused way. We all aimed to prove that the elections were fraudulent and that the resulting parliament would in effect be illegitimate. Our efforts were very successful among young Egyptians. The NDP "won" more than 95 percent of Parliament's seats. Everyone joked about it.

Yet I felt helpless. The regime was a hopeless case, and there seemed to be no way forward. I tried to use humor to relieve my stress. I remember writing many jokes about the rigged elections and one of them becoming a "top tweet": "Wikileaks would like to apologize for being unable to cover the Egyptian parliamentary elections of 2010, as its servers will not be able to withstand the number of forged documents expected to be revealed tomorrow."

We published a file that included all the photographs and video clips of the NDP's repressive and unabashed rigging. Everyone felt a mix of frustration and anger. The Muslim Brotherhood and the Al-Wafd Party decided to boycott the second round of voting. In a way, this was a personal victory for Dr. Mohamed ElBaradei, who was among the very first leaders to call the elections a ridiculous farce and demand a boycott. Developments proved he was right.

In December, a large group of computer engineers from Google visited Egypt and Jordan to hold two conferences with Arab web developers. The organizers asked me to deliver a presentation about the Internet in the Arab world. I spent quite a bit of time preparing the presentation, which urged Arab developers and media professionals to realize that they had a role as agents of change in the region.

In my presentation, I cited examples of entrepreneurs who utilized technology to create change. The young American of Bangladeshi origins Salman Khan was one. He was able to set up a simple YouTube channel called Khan Academy to facilitate basic education for 90 million people around the world. He uploaded videos of lessons in basic subjects that could be accessed by users anywhere, anytime. I also spoke of the Kiva initiative (Kiva.org), which mobilized $200 million in loans for 500,000 impoverished people in many countries. I closed the presentation with ten pieces of advice, the first of which was that every one of us can play a bigger role than we think is possible. "You are bigger than yourself," I said. Number ten was "Change the world! You can do it!"

In late December, AbdelRahman Mansour suggested that since the police celebrated National Police Day on January 25, then perhaps we should do something on that date. I thought his idea was brilliant. Many people loathe the police force. State Security officers and Central Security men were the most despised segment of society. The brutal and humiliating treatment of citizens by high-ranking officers was all too common. Yet we had no idea how to "celebrate" the day. It was quite challenging to come up with an event that would make a major statement without risking the safety of the participants. Should it be a demonstration, a silent stand, or something else?

I had an online brainstorming chat with Ahmed Maher, the cofounder of the April 6 Youth Movement, and it turned out that that group had done something similar before. On Police Day in 2010, they had called on activists to "celebrate" the police's transgressions against Egyptian citizens by protesting in front of the Supreme Court. We all saw the need for doing something unique in 2011. I recommended silent stands, in addition to an art campaign: a wall of fame that honored noble policemen, and a wall of shame that exposed the offenses of criminals in uniform.

I was getting excited about all the possibilities. We agreed to brainstorm further after the new year. Yet on December 30, I posted:

> January 25th is Police Day and it's a national holiday . . . I think the police have done enough this year to deserve a special celebration . . . What do you think?
>
> 471 Likes 119 Comments

5

A Preannounced Revolution

On December 17, 2010, Mohamed Bouazizi, an unlicensed vegetable-cart operator in Sidi Bouzid, a town 190 miles south of Tunis, had his cart confiscated by a policewoman, and when he complained to her, she allegedly slapped his face, humiliating him in front of everyone. He went to police headquarters to lodge a complaint, but the officers refused to see him. At 11:30 that morning he returned to headquarters and, as a protest, set himself on fire. He did not die immediately. He was transferred to a hospital near the capital. He passed away on January 4, 2011.

Protesters gathered at the police station the day after Bouazizi's self-immolation. They were met with violence and tear gas. Although the Tunisian state media downplayed the protests, the news spread on social networks and the protests spread and grew, reaching the capital by late December.

AbdelRahman Mansour wanted us to cover and support the Tunisian protests on *"Kullena Khaled Said,"* but at first I refused. I was worried that everyone would get excited and the Tunisian protesters would then be suppressed, leading to a new wave of frustration and hopelessness among our members. Yet AbdelRahman went ahead and published an image from the Tunisian protests, showing the protesters holding police helmets and sticks in victory. As I expected, strong disputes took place online between our members; it certainly wasn't time for us to appear impulsive. I reiterated to AbdelRahman my view that it was safer not to cover the Tunisian protests for the time being. Finally we agreed to delete that post and to wait and see how events unfolded.

The page's primary concern, after Khaled Said, had evolved to focus on the abuses of the Egyptian police. With every passing day I became more convinced that the police force was the chain that the regime tied around our necks; if the police force could be neutralized, the regime would be paralyzed. On the Facebook page we put a spotlight on the police's numerous violations and built up people's sentiment against them.

We tried as much as possible not to be indiscriminate in our attacks and not to make enemies of the entire police force. There were several times when we commended decent police officers who were true to their mission and performed their duties well. But unfortunately the transgressors were many, and with or without the Facebook page, the police were detested by most Egyptians.

The police force's official slogan had changed under the current minister of interior, Habib al-Adly, from "The police serve the people" to "The people and the police serve the nation." Even in its public relations, the police no longer served the people. Torture was not a matter of isolated incidents, it was systemic. The police force lacked the technology to help it solve crimes, so torture became one of its primary means of attaining information.

Most of the police's efforts were directed toward protecting the regime. The entire State Security apparatus, which was the strongest and most influential police division, shouldered a single mission: hunting down political opponents and individuals who showed signs of opposition. They terrorized them, threatened them, tortured them, and framed accusations against them. No wonder Egyptians referred to the police as "the government." Financial corruption was also widespread among police officers, mainly because their salaries were so low that some of them resorted to bribery and other illegal sources of income. Often when a traffic officer would pull me over to issue a ticket, I would bluntly say, "I do not pay bribes." Most of the time the officer would let me go, because he was either disappointed or embarrassed.

> A joke from somewhere: A police assistant's twin boys failed their exams. The first one entered to see his father, who beat him, cursed him, and the boy came out in a bad way. When the second one went in, his father looked at his grade report and said, "Just try to do better next year." When he came out, his brother was going crazy. "What did you do to keep him from beating you?" he asked. "I inserted a five-pound note inside the grade report.
>
> 1,033 Likes 187 Comments

I wanted to end the year on a positive note, despite everything depressing that had taken place. On the final day of 2010 I started to publish positive messages from members and asked for more to be sent in, to inspire hope as we welcomed the new year.

> Happy New Year . . . Hopefully next year would better than the last . . . Let us all encourage one another to be positive and proactive . . . I ask everyone to tell us what thing he or she did during 2010 that made them the most proud.
>
> 1,057 Likes 477 Comments

> I will start the list by sharing the positive things members did and mentioned in their comments . . . Let our actions be a source of inspiration to others . . . We'll make our country a better place . . . As long as we are together . . . defending the oppressed and the weak . . . We will be stronger . . . *Kullena Khaled Said.*
>
> 120 Likes 14 Comments

Many messages flooded in, along these lines:

Eman Farid: I'm grateful I did many things that make me proud. First, I passed all my classes at the university for the first time and scored a grade average of "Good." Second, I participated in a Silent Stand for the first time, and I was alone, and a police guard chased after me but nothing happened, thank God. Third, I went to vote for the first time and witnessed the rigging firsthand. Oh God, please turn all Egyptians into proactive citizens who change themselves first before they decide to try and change others!

155 Likes 27 Comments

Khaled Mansour: Thank you to the admin of *"Kullena Khaled Said"* for directing our anger toward something positive . . . The idea is how to manage our anger . . . Our anger will make us insist on fighting injustice and attempt to address all problems with positive action. I will not direct my anger at myself again.

132 Likes 5 Comments

We were quite busy trying our best to end the year in a spirit of optimism, completely unaware of the calamity that was about to befall the nation. Only minutes into the New Year, a bomb exploded outside the Two Saints (Al-Qiddisayn) Church in Alexandria while a service was in progress and devout Coptic Egyptians filled the building. The church trembled, and the sound of the explosion was almost deafening. Many people inside fled from the building, thinking that an earthquake had struck the city. Instead, they came out to find the remains of twenty-one people lying on the ground, body parts scattered all over. Dozens more people had been injured. The scene was dramatic: flames were blazing, blood was everywhere to be seen, and many were just standing there in horror. The images and videos that people took on their mobile devices that night shocked the nation.

From the beginning, our Facebook page had steered away from discussing sectarian incidents, to avoid the eruption of warring comments between members. Given a handful of extremists on both sides, open forums could easily descend into shouting matches. The regime's policies, whether intentionally or unintentionally, had engendered a sharp divide between Muslims and Christians, in spite of the fact that generations of Muslims and Coptic Christians had lived together peacefully in the past. The regime was good at utilizing this divide to create a perception that without Mubarak in power, Egyptians would break out into sectarian warfare. As a result, Mubarak managed to market his police state successfully to the international community as the lesser of two evils.

Whenever violence between Muslims and Christians occurs in Egypt, it usually starts for nonreligious reasons. Sectarian strife, particularly in the nation's underdeveloped cities and villages, primarily results from interreligious commercial or emotional relationships that have gone bad. As a result of poor education and the absence of a just system, predicaments that would normally be sorted out in civil courts turn into battles between family clans. Occasionally fighting will erupt between Muslim and Christian families because of a single business transaction. Moreover, crimes that coincidentally involve people of both religions are sometimes perceived as deliberate attacks on the victims because of their religion, which leads to retaliation by a victim's family. Sometimes a girl decides to switch faiths and elopes with a young man from the religion she has recently embraced because her family will not authorize the marriage. Then her family announces that their daughter has been kidnapped and forced to renounce her faith. Innocent love quickly turns into cycles of vengeance, and cases of this sort usually lead to huge debates across the country, resulting in more frustration and anger on both sides.

After the church bombing, I searched online for an image of the crescent and the cross together to try to soften the blow to our national fabric. I asked everyone on our Facebook page to change his or her profile picture to this image, on which I wrote, "I am an Egyptian against terrorism."

I wish all of us could replace our personal profile photos with this one, even if just for a day . . . We must defuse the strife.

516 Likes 237 Comments 64,532 Views

The first thing we did was to publish painful images and video clips of the bombing. Some page members accused us of adding fuel to the fire. Our reason was clear: knowing the facts was important. I posted:

Posting the images aims to present the facts, not to further ignite the problem . . . This page was only created as a result of Khaled Said's image after he was tortured . . . If we had not seen it, we would not have cared . . . We must all stand together now and forget our differences. We must all fight the criminal who did this regardless of who he is or who he works for.

218 Likes 94 Comments

Egyptian media usually shied away from publishing images of attacks on Copts or churches. The regime feared increasing the strife, in case it went out of control. Yet with the spread of the Internet and satellite television, Coptic Egyptians' emotions would be even more strongly aroused by gruesome images circulated within their communities while Muslims remained unaware of the atrocities because they never saw them. In order to create empathy for the victims, it was critical to publish all the video clips, including the bloodiest ones, for all our members to see.

I urged page members to donate blood and to try to remain calm and rational. Igniting sectarian strife is exactly what the perpetrators want, I told everyone. A member of the page wrote a comment that any fighting or anger that erupted between Muslims and Christians at this point would be nothing short of congratulations for the culprits. He advised members to channel their anger into rational thinking, so we could devise practical steps toward real change that would eradicate this problem at its roots. The message found great acceptance among page members.

Many religious leaders from both the Muslim and Christian faiths were urging people to calm down. I posted an important interview with Moez Masoud, a young religious figure known for his intellectual and contemporary interpretations of Islam. Moez, whom I had known for years, clearly stated that Christians were as integral to the fabric of Egypt as Muslims, and that it was crucial for Egyptians to unite and not to give the terrorists what they desired.

> "On January 1, all of humanity died in Alexandria . . . If we want to reclaim Egypt once again, we must rebuild our nation; rebuild industries, the economy, politics, and set forth a real renaissance." Moez Masoud is saying some critical things about the recent events. We can change our country to the better with our own hands, and we will!
>
> 332 Likes 85 Comments 74,690 Views

Infuriated Christians began protesting across the nation. They were extremely worried that the perpetrators of this atrocity would get away with their crime, something that had happened in the past. The protesters' anger mounted, and they started to block major roads in Cairo. A protest in Shubra, a very busy Cairo neighborhood with a large number of Christian residents, turned violent. Clashes between protesters and police left casualties on both sides, but mainly among the protesters. One video that spread widely on the Internet showed a soldier repeatedly beating a protester on the head with a stick, despite the protester's blood and screams.

The Ministry of Interior issued a report condemning the "violent" protesters for terrifying the residents of some Cairo neighborhoods and stating that two officers and twelve soldiers had been injured. The ministry did not even mention the casualties on the protesters' side. Eight Muslim political activists who took part in the protests were arrested by the police, apparently as a message to Muslims not to join together with Christians. The regime seemingly wanted the events to be perceived as exclusively sectarian.

In an effort to ease the tension, our page called for a stand to take place on the Orthodox Christmas Day. Orthodox Christians make up the majority of Egyptian Christians, and they celebrate Christmas on January 7. Many people feared that another Egyptian church would be targeted somewhere during the Christmas celebrations. This is when the prominent Egyptian engineer and intellectual Mohamed al-Sawy suggested that Muslims form human shields around these churches, to protect their Christian brothers and sisters during prayer. It was a message that terrorism would not defeat our unity. On the page, I wrote that the perpetrator did not just cause twenty-one Christians to die, he also caused 80 million Egyptians to fight.

The January 7 stand was covered by *Al-Shorouk* on its first page and also by prominent news websites. Mona al-Shazly also covered it, and several public figures promoted participation. Our Facebook invitation reached 170,000 people, of whom 18,000 confirmed participation. The idea of protecting the churches was also a success and also received considerable media attention. In practice, however, turnout was significantly low for both, largely because of violent protests led by angry Christians in popular neighborhoods.

Nonetheless, the events provided the first time that many Muslims had ever entered a church in Egypt. Even some women in the *niqab*, the face veil, attended church services that night. We published their images, and those of Muslims who held up signs stressing the unity of Egyptians in the face of terrorism in all its forms.

While our nation was recovering from the incident, a forty-second video of the corpse of a bearded man in his thirties spread online. The clip made me recall what had happened to Khaled Said. This man's name was Sayyid Bilal. He was one of the detainees suspected of taking part in planning and executing the Alexandria church bombing. He had not been able to endure the police's torture and had died because of it. The police used their traditional methods, which included applying electric shocks to sensitive areas of the body. Sayyid Bilal's family had taken the photo of his corpse and published it online. Our page immediately adopted Sayyid's case, and to show that we did not discriminate between Muslim and Christian, we published a joint picture of Sayyid Bilal and Martina Fekry, one of the Christians who had died in the church bombing, and noted that they had both been victimized for crimes they had not committed.

> Maryam Fekry and Martina Fekry . . . victims of the criminal attack . . . in a photo with their friends . . . God bless your souls, Maryam and Martina . . . Truly, terrorism is godless.
>
> 856 Likes 565 Comments 86,716 Views

> From journalist Abdul Moneim Mahmoud . . . Sayyid Bilal's family was pressured to prevent them from pressing charges . . . We must break the security pressure . . . Take your mother and wife to express condolences and support them not to give up their case . . . Bazar St., Namous Bridge . . . Near Zahiriya Station and Fouly Mosque . . . Send a message to Sayyid Bilal's brother-in-law and lawyer Izba and ask him not to give in to security pressure by dropping the murder charges against State Security.
>
> 538 Likes 121 Comments 41,926 Views

Sayyid Bilal came from a poor family and had a small child. The page gave a lot of attention to his case because it addressed the kind of violations facilitated by the Egyptian emergency law and the State Security torture practices against activists from religious groups. AbdelRahman Mansour was able to coordinate with the *Al-Dustour* newspaper reporter Abdul Moneim Mahmoud, who was in constant touch with Sayyid Bilal's family, and we published a picture of the man's young son and another of his mourning mother. The regime was getting more exposed in the eyes of the public. Many Christians wrote posts expressing sympathy for Sayyid Bilal's family.

In an attempt to renew our optimism, on January 7 I launched a campaign called "The Week of Goodwill for Khaled Said." The basic idea was to ask people to do something positive in the society and dedicate it to Khaled.

As the page dealt with day-to-day events and promoted goodwill, there was little time to look ahead. Yet with January 25 only a little over two weeks away, it was time to put together a plan, democratically.

> We need ideas for Police Day on January 25th: Because these people work very hard at humiliating, torturing, and sometimes killing Egyptian citizens, we should not pass up the opportunity on this day to let them know we will not forget . . . Please, everyone with an idea should send it forth. Let's make them unexpected and different ideas. Hopefully we will be able to repay their kindness.
>
> 506 Likes 272 Comments 183,015 Views

To my dismay, page members were not discussing National Police Day as vigorously as I had hoped, though several of us were tossing ideas back and forth. But history kept intervening—until January 13, when events outside Egypt suddenly gave us the spark we needed.

On that day, Tunisian president Zine El Abidine Ben Ali, who sensed the citizens' blazing anger against him after a speech in which he described the protesters as masked gangs serving foreign conspiracies, came out with an unprecedented statement for an Arab dictator. He addressed the people with a phrase that has become famous: "Now I understand you." Ben Ali told the Tunisians that he understood that they were frustrated by the performance of his government. He appreciated that they would no longer tolerate humiliation. It was the speech of a defeated man.

Ben Ali's speech changed everything. This was the perfect time to start discussing the developments in Tunisia on our page. The victory of the people of Tunisia would send a strong message to the Egyptian regime and, more important, to our Facebook page members: we can effect change in Egypt.

As usual, I decided to solicit members' opinions and ask what they thought about the idea of covering the events in Tunisia. Those in favor made up a sweeping 86 percent. Tunisia's example was an inspiration for all Arabs. We were starting to feel that the impossible was no longer impossible, and I posted:

> There is no president or government or police force stronger than the people . . . The day before yesterday the Tunisian president told his people that the protesters were gangsters and terrorists who will suffer severe punishment . . . And today, after tens of thousands of Tunisians took to the streets, he felt the danger and sacked the minister of interior and released all detained protesters.
>
> 348 Likes 76 Comments 99,472 Views

I had never imagined when the Tunisian demonstrations began in December that the regime might ever surrender, let alone in a matter of weeks. Yet on the day following Ben Ali's speech, the Ministry of Interior was cordoned off by protesters and Ben Ali fled to Saudi Arabia. On the page, I posted Moataz Kattan's design of the Egyptian flag with the Tunisian red crescent and star to the side. It was a sign of our solidarity with Tunisia and our hope that Egypt would follow the same course.

> For the first time since the day I was born I saw an Arab president pleading with his people . . . apologizing . . . saying, "Now I understand you" . . . "I am sorry" . . . "I have made a mistake" . . . "I will not amend the constitution to reelect myself" . . . "I will reduce prices and offer grants to the unemployed" . . . "I will lift the ban on the media" . . . "I will allow for real political pluralism" . . . "Please . . . please . . . I am sorry" . . . O youth of Egypt, listen so you know that no government is stronger than its people . . . A strong salutation of respect to the free people of Tunisia!
>
> 793 Likes 197 Comments 26,128 Views

The comments on the page were powerful. Some demanded that we call for our own uprising. Others mocked the passivity of Egyptians. Many people launched self-ridiculing jokes, as we Egyptians love to do. Some of them attacked me personally—for example, "If Bouazizi had burned himself in Egypt, the *'Kullena Khaled Said'* admin would have organized a silent stand."

I wrote my first message containing hints about Hosni Mubarak:

> Goodbye, Zine El Abidine the Runaway . . . Enter into history's garbage disposal . . . But please, don't anyone place the lid back on!
>
> 1,105 Likes 259 Comments 94,284Views

By January 14, I started to believe that we could be the second Arab nation to rid itself of its dictator. Egypt's political, economic, and social conditions were worse than Tunisia's, and the level of anger on the street was much greater. The only thing that separated Egyptians from a revolution was our lack of self-confidence and our exaggerated perception of the regime's strength. Yet after what happened in Tunisia, I thought the Egyptian masses might finally get the message and break the psychological barrier of fear.

Egypt occupies a unique position in the Middle East. Egyptians proudly consider themselves to be the cultural and scientific leaders of the Arab world. Our pride had now been challenged: Tunisia had taken the lead in the quest for liberty. It was a shot in the arm. The psychology of the proud and courageous Egyptian played a major role in enabling our country to follow in Tunisia's footsteps.

With time, young Egyptians were getting angrier. Analogies between Tunisia and Egypt were increasingly being drawn. Some page members were even openly requesting that *"Kullena Khaled Said"* call for a revolution. I thought I would test the waters:

Today is the 14th . . . January 25 is Police Day and it's a national holiday . . . If 100,000 take to the streets, no one can stop us . . . I wonder if we can??

3,022 Likes 1,748 Comments 176,013 Views

My post received more than three hundred comments in a very short time. Some members had been hoping for just such an invitation. Others were negative, convinced that Egypt would never change. And others called my post an invitation to chaos that would take the nation into a dark tunnel. I read every single comment with great attention.

Ahmed Moatamed: It's a great idea but 10 days give us little time to prepare for something as such. There needs to be coordination with other movements and political forces, and people who wish to see Egypt change. May God help us and may the people wake up.

Mohamed Zayed: That would be greeeeeeeeeeeeat. I will be the first person at the location you specify.

Mostafa Kamel: It's a great idea but it needs careful organization.

M. Sherif: We can. We can. We can. And 50,000 in Alexandria too.

Hossam Khafagy: No, we can't . . . because the whole country is on holiday except the police . . . they sleep standing, God bless them.

Hamdy: You are really a terrorist.

Khaled Bahgat: Hey, admin, will we still wait? Let's do it now when the people are excited.

Yara Ayman: Again with your stupid stands? Enough is enough. Do something useful for a change.

Sameh el-Balah: Let's start rallying the masses now . . . We can do it, God willing . . . Religions say we can . . . And history says we can!

Mohamed el-Amry: No one will do anything and you'll see. All we do is post on Facebook. We are the Facebook generation. Period.

Moody Basha: Now you're playing serious, Mr. Administrator . . . You seem to be on fire . . . But look to the future and think what could happen. Egypt is not like Tunisia, and the army won't go against Mubarak, believe me . . . But you know what? I'm with you and I don't care what happens.

Ahmed el-Omda: The government is controlling the country with a fist of steel. I don't think we can do what Tunisians did. Once upon a time Egypt was a model for Arab people everywhere. If only we could turn back time, to the Oraby and the July revolutions.

Michael Salah: We can if we put our brains to work.

Two hours later, AbdelRahman came online. I quickly asked him, "What if one hundred thousand people took to the streets on Police Day?"

"Can we do it?" AbdelRahman asked.

"To be honest, I don't know," I answered, "but I feel that the level of energy is soaring. However, we know how afraid the Egyptians can be." Then I shared my doubts with him. I told him I was not a rebel. I was not crazy about the idea of persuading people to try something with such unknown consequences. I also did not have experience in managing such a crowd if this were to happen. I asked AbdelRahman if it would be better for a strong political opposition movement to call for such an event so that we could just rally for it. His response was that there was no strong political opposition in Egypt. He was right.

I was quite optimistic, but not without caution. I was afraid such an event would be a spectacular failure and our generation would receive the shock of its life. But I began with a campaign to counter the negative voices.

I'm sure that months ago some users on Tunisian Facebook pages were passive discouraging voices who said, "There's no hope and the country will never change; the Internet is monitored and the Ministry of Interior will arrest you." . . . These people are everywhere at all times, but poor them, they become so ashamed once victory is attained . . . God willing, we will reclaim all our rights and we will act as one.

489 Likes 78 Comments 99,149 Views

I also started posting messages and comments by members as a form of encouragement to others to participate and contribute positive opinions. One of the opposition newspapers published an abstract from a new book called *The Road to the Presidential Palace,* by Mohamed Ali. In it he mentioned that if 100,000 citizens gathered in Tahrir Square, they would get what they wanted. I wrote a message on the page to the people of Egypt—"Let January 25 be the torch of change for our nation"—and I published the writer's statement.

In late December, after AbdelRahman's suggestion to celebrate National Police Day on the twenty-fifth of January, I had initially named the event "Celebrating Egyptian Police Day—January 25." Yet now, less than three weeks later, with the toppling of the Tunisian president and the mounting anger of many Egyptians who sought to replicate this situation with Mubarak, it became necessary to completely reposition the event. I found myself unable to resist the word *revolution.* Every time I attempted to steer away from it in my thoughts, it kept coming back. The decaying regime had become Egypt's main problem, and the only way forward was to remove it. This was ironic, given that I had clearly stated on more than one occasion that I was not a revolutionary.

Although it took no more than a few keyboard strokes and a single mouse click to change the event's name to "January 25: Revolution Against Torture, Poverty, Corruption, and Unemployment," my mindset changed drastically immediately after I did so. I felt an adrenaline rush, only this time it wasn't fight or flight, it was fight or fight. After posting the event's new name, I was ready to face any and all consequences.

I deliberately included poverty, corruption, and unemployment in the title because we needed to have everyone join forces: workers, human rights activists, government employees, and others who had grown tired of the regime's policies. If the invitation to take to the streets had been based solely on human rights, then only a certain segment of Egyptian society would have participated.

> After all that's happened in Tunisia, my position has changed. Hopes for real political change in Egypt are much higher now. And all we need is a large number of people who are ready to fight for it. Our voices must be not only loud but deafening. I swear to God, I'm going to participate in Jan25 and I'm ready to die to free Egypt from tyranny.
>
> 410 Likes 91 Comments 101,422 Views

The response was immediate and positive. AbdelRahman Mansour was only two days away from starting a forty-day stint at a closed training camp as part of his compulsory military service. I remember joking that by the time he came back, Mubarak would have stepped down and our page would have reached one million members. Little did I know that these predictions would actually come true.

I changed the profile photo on the page to an Egyptian flag containing a Tunisian symbol in its red section. In the black section, I wrote, clearly addressing the regime, "See you on the 25th of January."

In response to one defeatist commenter, I wrote:

> Don't give up, Mostafa . . . Believe me, the Egyptians are not cowards . . . But each person is afraid they will be alone when they rise up in anger . . . Egypt has three million

unemployed citizens . . . If we all act, no one can stop us. And believe me, the army will take our side if we strategize correctly.

302 Likes 155 Comments 96,919 Views

The official Egyptian press played down the story of Ben Ali's resignation. It was pitiful. The front page of one of the largest state-owned newspapers on January 15 carried a headline about his "departure" but overshadowed it with much larger type for this: "Egypt Soars High: Mubarak Achieves the Highest Levels of Economic Security for His Country." Everyone could see that the regime was terrified of the events in Tunisia. The paper refused to use the word *revolution.* I published a scan of the newspaper on the Facebook page, and it quickly became the laughingstock of Egyptian cyberspace. Just about everything the regime did ended up having the opposite effect from the one intended in our virtual world.

On January 16 the minister of foreign affairs dismissed the possibility that Egypt would follow Tunisia's lead as "nonsense." He added that each society had its own unique circumstances, and stated, "As for people who have let their imaginations run wild and spilled a lot of oil over nothing, their clothes will be stained by the oil. There are media outlets that are out to ignite fear in Arab societies and are hoping to destroy them. Unfortunately, they are all foreign satellite channels."

During the week that followed the Tunisian revolution, more than five Egyptian citizens tried to set fire to themselves outside the Parliament building. The first case of self-immolation happened on January 17 and was covered intensively by the media. The head of Parliament issued statements to try to calm everyone down, and the minister of health visited the person, who had not died, at the hospital. The police even prevented gas stations from selling gasoline in containers.

I started rallying everyone for January 25 by appealing to organized groups. I began with the "ultras," as zealous soccer fans are known – many thousands in number.

> To the ultras of Ahly, Zamalek, Ismaili, and Itihad soccer teams . . . If you exert the same effort you do for any soccer match on the 25th of January, you will help Egypt change . . . Let us all be ultras of Egypt . . . Let us all take action and take to the streets . . . Who among us is an ultras member and prepared to cheer for Egypt?
>
> 1,616 Likes 408 Comments 207,557 Views

I was careful not to portray January 25 as a *"Kullena Khaled Said"* event. This made it easier for several other pages and political parties to promote the invitation as their own. Among the first of these Facebook pages were the April 6 Youth Movement (80,000 members), the Nizar Qabany page, named after the late poet (157,000 members), and the Egyptian Sugar Cane Juice page, which had 250,000 members. Even humor pages were now promoting January 25! The avalanche had started.

We were promptly accused of treachery, foreign agency, and vandalism. I tracked the accounts of our attackers to discover whether they were real accounts or fake ones created by the NDP's Electronic Committee. Sure enough, there were hundreds of recently created bogus accounts. I banned them all and posted screen shots of a few to expose them.

Activists had only a few days left to do everything we could to encourage all Egyptians to take to the streets. It was a revolution without a leader and without an organizing body. All I knew was that if hundreds of thousands of Egyptians participated, the regime would have to respond. I did not coordinate extensively with other pages to get them to publish our content, nor did I try to get credit for our page. The idea was that everyone had a right to advertise this day. Spontaneously, however, everyone did rely on the main event page posted on *"Kullena Khaled Said"* to promote the January

25 revolution, which came to be known as "Jan25" in the virtual world.

I prepared a few guidelines to constantly keep in mind:

1. Reveal the regime's weakness and fear.
2. Draw parallels to Tunisia.
3. Report on the series of Tunisian victories that followed Ben Ali's escape.
4. Expose all of the regime's economic, political, social, and security failures.
5. Invite others to promote Jan25 through their writing, poetry, songs, and designs.
6. Assure everyone that the army's position will be honorable and that it will take the side of the people.

This strategy was largely successful, primarily because our pompous regime never took seriously anything the people did. The statements released by the government, one after another, just added fuel to our fire.

Meanwhile, developments in Tunisia were very positive. Police officers who had been implicated in violence were arrested. The Tunisian army had acted honorably during that country's revolution. They refused to attack the people or even point their weapons at them. We had to achieve a similar understanding with our own army. I began to publish any and all photographs that depicted a sympathetic and emotional interaction between the Tunisian people and the Tunisian army. The reaction was spectacular.

> The Tunisian army set an example of patriotism . . . An army where officers and soldiers learn to defend their homeland and their countrymen cannot possibly fire at their own people and kill them . . . During the funeral procession of one Tunisian martyr killed by the police's bullets, one of the army officers stood in respect for the martyr.
>
> 773 Likes 231 Comments 191,658 Views

One of our Egyptian army officers was quick to respond to our posts. He sent a message to the page's e-mail account from a generic address with an image of his uniform and a sign that said, "Dear Egypt, victory is soon. Jan25 is the day." I replied to him and asked for his identity, but he adamantly refused to reveal it, saying that doing so would put him at great risk. He told me that the general mood within the army's ranks was supportive of our revolution. The picture was a major boost to our members' spirits.

Images of the Tunisian demonstrations were magnificent: hundreds of thousands of united citizens carried nothing but the Tunisian flag and signs that demanded that Ben Ali step down as president. We were even more inspired by the images of the Tunisians' ultimate victory. One of the most expressive images featured a soldier's cap in reverse position, denoting defeat, against a backdrop of protesters. I commented on this image: "Nothing rises above the voice of the people."

> To everyone who cried in tears on the day Egypt lost at the World Cup (and I was one of them), we must now cry in tears that Tunisia gained the Cup of Liberty . . . Liberty is much more important than a soccer match . . . Dignity and humane treatment are much more important than the World Cup . . . We must reclaim our rights and this is why we must all take action on Jan25.
>
> 448 Likes 76 Comments 107,593 Views

I had become fully devoted to the Facebook page once again. I went to work with nothing on my mind but what I must post on the page. Every two or three minutes I refreshed the page to read all the new comments. My daily share of sleep was barely three hours. I spent all my time at home inside my room.

I published an image of Mubarak and Ben Ali with a caption under Mubarak saying, "You have led and we shall follow." Later, one of the members wrote a satirical letter of resignation in Mubarak's voice.

I contrasted positive images of the Tunisian revolution with negative ones of Egypt. I reposted the torture videos, now that the page's membership had reached 365,000.

The campaign was not limited to *"Kullena Khaled Said"*; other pages, especially that of the April 6 Youth Movement, worked in parallel, without any coordination of effort. Yet our independent work was all aimed at one thing: Jan25 was not going to be like any other protest the regime had experienced.

> We will demand our rights on Jan25 . . . Let us focus from now until Jan25 on discussing Egypt's economic condition and our living standards . . . We must reach out to the helpless layman who only cares about finding his loaf of bread . . . Let's refrain from elitist sophisticated talk so we don't end up only 1,000 or 2,000 on the street . . . The Tunisian youth began their demands with solving unemployment and inflation . . . And when the government was not responsive they acted . . . We must do the same.
>
> 322 Likes 151 Comments 119,239 Views

Mention of Mubarak had been off-limits on the page. But as soon as Ben Ali fled Tunisia, this was no longer the case. Earlier I was anxious not to lose the support of the silent majority; I wanted to remain in touch with the Egyptian masses and not attract accusations of political motives. Now everything was different. I began to criticize Mubarak and his autocratic practices directly.

An anonymous designer posted a brilliant image that compared American presidents from 1981 to the present with Egypt's presidents of the same period. Across from Reagan (1981), George H. W. Bush (1989), Clinton (1993), George W. Bush (2001), and finally Obama (2009), the Egyptian side offered nothing but various photographs of the same man. Mubarak's images over thirty years showed clear change; his smile died out and his facial features aged. It was a simple and expressive design that showed beyond doubt that we had been ruled by a dictator for three decades.

Many creative ideas that I responded to came through the members of the page. One of the best ones was to distribute mass text messages (SMS's) that publicized Jan25. It was crucial for the invitation to reach all Egyptians, even if they did not participate, so that they anticipated the event and took the time to determine their positions. Reaching working-class Egyptians was not going to happen through the Internet and Facebook. Youth groups that had mobilized through the Internet printed fliers of the Jan25 invitation and distributed them, together with the SMS messages.

> I never saw this on Facebook before . . . The Jan25 invitation reached 500,000 Facebook users . . . 27,000 have RSVPed . . . The important thing now is to spread out to streets, factories, mosques, and churches . . . It is crucial for the massive [working class] districts of Shubra, Boulak, Manshiyet Naser, Ein Shams, and Mahalla to hear about us . . . People in the villages must know there is a solution . . . That we will all take action and say "No" . . . That we demand our rights . . . Let's do this, Egyptians . . . Let's show the world that we are not cowards and that we are ready to sacrifice anything for our rights.
>
> 559 Likes 239 Comments 118,928 Views

The event page was inspiring. During the most successful of the Khaled Said stands, our invitations never reached more than 100,000 people. In just two days, the Jan25 event reached half a million Egyptians online, and 27,000 confirmed attendance. This would be no ordinary demonstration.

I asked all page members to promote the revolution at forums, on Facebook pages, and in other famous online communities. Their response was significant. Jan25 became a recognized date for most Internet users in Egypt. I also posted a few suggestions on the page to deliver the invitation to the largest number of people possible, by sharing links to the event page, sending text messages to acquaintances, printing fliers, and asking each member to invite five other people in person to come out on Jan25.

There was no actual plan for the day, at least not yet. I did not know what we were going to do, exactly, and therefore I decided to write to Mostafa al-Nagar, the activist leading Dr. ElBaradei's campaign. I also wrote to Ahmed Maher, the April 6 Youth Movement coordinator, whom I had met only once and who did not know that I was the *"Kullena Khaled Said"* administrator. I sent them both an e-mail message saying that we needed to start planning for Jan25 and that they must take responsibility for coordinating the fieldwork with all the other activists and inform me of suggested locations.

Mahmoud Samy, the former administrator of "My Name Is Khaled Mohamed Said," sent me a message on January 15 suggesting that we cooperate: "We need to coordinate for Jan25. I'm almost sure you will not respond for the reason we both know, that you decided to stop dealing with me since I discovered your name . . . But are you or are you not ready, to coordinate with April 6 and others? If you are willing, then I could handle this for you on the ground, or I could have someone I trust take care of it, if you still deem it problematic to deal with me directly . . . It is time for the Egyptian flame to ignite."

Ahmed Maher suggested that demonstrations begin from the Ministry of Interior or in Tahrir Square. I opposed this idea strongly. To me, these locations typified traditional demonstration venues, ones that were more commercial than residential and where the security apparatus was most experienced in controlling protesters. I argued that we should begin in popular neighborhoods.

I also spoke with Mahmoud Samy about locations. He was very excited about Arab League Street. This is a major street that he thought was ideal because it is quite long and runs through several popular neighborhoods. Its breadth and length would also make it difficult for the security forces to seal off. I knew I was not a field expert and so I deferred to Mahmoud Samy and Ahmed Maher, among other activists, and asked them to coordinate with each other. Of course I also asked Samy not to reveal my identity to Maher.

On January 17, I had my last chat with AbdelRahman Mansour before he began his compulsory military service. It was a very emotional conversation. He said he would miss the page more than anything else. He expressed his frustration at potentially missing out on what he had been dreaming about for years: the beginning of real change in Egypt. He asked me to continue reminding the page's members that Jan25 would only be a beginning, so that no one would give up if the turnout was low. It was amazing how AbdelRahman and I had worked well together for months without any face-to-face interaction. The conversation became very emotional as we both said goodbye.

Meanwhile, Maher sent me the news that activists had met with key leaders of the ultras from Ahly and Zamalek and that they would participate. He added, "By the way, I don't know your name, but I swear that I really respect you."

I wrote back, "I respect you too, but you don't have to know my name. Call me Khaled. I want nothing other than for my country to change. I don't want my name to become known, not before Egypt changes and not after. I really wish we could get this change out of the way because I need to focus on my personal life. Come on, let's keep up the great work. It was a great idea to meet the ultras. This is what I call masterminding." Later I sent him an online message suggesting that activists take to the streets in gradual numbers.

I mentioned on the page that we would not announce the time and place for the Jan25 demonstrations until two or three days before the event. We wanted to make it hard for security forces to prepare themselves. Some members mocked the idea of a "revolution" that was predetermined in terms of time and place. The way I saw it, however, bringing together large groups of people that were hard to control could potentially lead to the revolution for which we were so hungry.

Despite my optimism, I began worrying about the turnout. The pro-revolution page members were developing sky-high expectations, and low turnout would have been fatal for the involvement of many of them. I remember telling some of the activists with whom I was anonymously coordinating, "I'm counting on Jan25, but if, God forbid, the day fails, I will close the Facebook page, because that will be the end, at least for me."

I wrote a note on the page under the title "Jan25—Important Points to Highlight; Our Reasons for Protesting and Our Demands for the Government."

> No one would have ever imagined that Police Day would turn into a day on which Egyptians would wake up, revolt, and demand their rights, and that the invitation would reach 700,000 people on Facebook . . . Up to now, there has been no telling whether we have succeeded or failed, because our struggle has only been electronic so far and limited to comments and "likes" . . . But what I know for sure is that there has been a response that I have never before witnessed with any of the Khaled Said stands. Everyone is working and hoping Jan25 will be a success, because everyone is fed up and suffering under our current circumstances.
>
> Egypt is full of problems, and our aim was never to increase them through demonstrating . . . Yet our main problem is that the government and president do not listen to their people . . . They are used to the practice of "the president commands and everyone obeys," and people who object get arrested and beaten and humiliated, or, if their voices have no real influence, they are left to object as much as they want . . . This was the same practice of the Tunisian president and his government. Even when Bouazizi set himself on fire, the government's reaction was minimal and negative. They did not expect the people to continue or the number of citizens demanding their rights to increase. Yet eventually the Tunisians' president vowed not to run in 2014; next he said the country would hold elections in six months, and finally he ran away like a criminal.
>
> Our main problem in Egypt is, ironically, that every one of us only sees his or her problems . . . We don't realize that even though our troubles may be different, the symptoms are one and the causes are one . . . We ceased to feel for the poor . . . How many of us have thought about how a person can live for a whole month on a salary of 300 pounds? We all ignore these realities and brush them off as irrelevant to us . . . What happens to such a person? He or she will steal, accept bribes, and open a door to corruption. By the way, this person could affect you directly because he is the manager at a government

agency, he teaches your children, he bakes your bread, etc., etc. . . . And if this person decides to be an honest citizen who does not accept bribes, then he or she will end up living a life of subhuman standards, losing all sense of belonging and patriotism, and will then pass on these feelings to his or her children . . . And this is where the poor people's statement of "This is not our country, it's theirs" comes from ("theirs" referring to the regime and the rich businessmen around it).

Religions urge humans to be concerned for those around them . . . A society's problems are much more dangerous than an individual's problems . . . We all live in one boat, and if we leave someone to keep punching holes in it, our boat will eventually sink . . . And none of us will survive.

In the past, the government maintained a long-standing tradition of telling lies to the people through its state-owned puppet media . . . However, this has changed during the past few years . . . and people are really starting to see and hear about the nation's problems and their real magnitude . . . And the government's reaction was always either pitiful or ridiculous . . . The people lost trust in the government, and the government thinks that since the people are silent, then they are satisfied, or that there is no problem until the people take action.

Jan25 is the beginning . . . A beginning for what, exactly? A beginning for us to join forces and start to apply pressure . . . And to have specific demands and a legitimacy that most Egyptians would agree about . . . And for these demands to resonate, we will all take to the streets in the form of sit-ins, protests, and marches in all of Egypt . . . The objective is not to overthrow the regime or to change the president overnight . . . Because the problem now is not the president . . . The problem now is an entire system that needs to change, and the chances of changing it are tied to the necessity that we all change and demand change, and apply pressure and reclaim our rights.

Jan25 must be a day for the common striving citizen who is unable to make a living . . . who works 18 hours to pay for his daughter's chemistry lesson . . . or who stands 3 hours in line for bread . . . or who saves money with a group of friends to be able to buy meat . . . Even though I am none of these people, my conscience stipulates that I help every one of them . . . And it is not only my conscience, but my mind too . . . Because if this condition persists in Egypt, we will soon suffer incredible societal problems and people will start killing one another and people will become bitter in treating one another . . . My message to all of you is that we focus on the messages that reach all people . . . Not everyone appreciates the meanings of freedom or cares about democracy.

I am taking to the street on Jan25 . . . There is no way I will react to violence with violence . . . There is no way I will strike back at anyone who strikes at me . . . But I will defend myself and protect myself . . . And I am ready to die a martyr, because it is necessary for our country to change, that people who take to the streets be willing to sacrifice for their nation . . . Egypt will not change through Facebook . . . But Facebook can help us discover the news and unravel the truth so we can take action in reality.

I hope my message has reached you.

And I hope you forgive me for its length.

603 Likes 270 Comments 140,198 Views

I coordinated with Mahmoud Samy, Ahmed Maher, and Mostafa al-Nagar by e-mail from Dubai. Only Mostafa and Mahmoud knew that I lived in the UAE. As fate would have it, however, Maher and I accidentally met right before Jan25.

An event organizer working with Al Jazeera news channel had called me in December to see if I would speak at a conference to be held in Doha, Qatar, about the Internet in the Arab world. I liked the idea of visiting Qatar for the first time in my life. I frequently spoke at conferences because of my Internet experience and my work at Google. This one happened to be scheduled for January 19, only five days after I had announced the Jan25 protests.

I thought of canceling the trip and focusing on the Facebook page, but I decided not to amend my plans, because I hate last-minute cancellations. I decided to attend the conference but to spend all the time before and after delivering my presentation working on the page in my hotel room. As soon as I arrived, I discovered that Ahmed

Maher and Israa Abdel Fattah, the young activist who had founded the Facebook group that promoted the call for the April 6 strikes in 2008, had both been invited to the conference, since they were both active bloggers. It was a strange coincidence.

I met them briefly before the first conference session began. At lunchtime, after delivering my presentation, I sat at a table full of Egyptians. I asked Ahmed Maher, as someone who followed events closely, what he thought of Jan25. He said that he had high hopes and that the day would be bigger than the strike of April 6. I asked if the locations had been determined, and Israa responded that the *"Kullena Khaled Said"* administrator was coordinating with political activists and that locations would be announced soon. It was a surreal conversation. The funny thing is that after lunch I sat in a corner where I could still see Maher and sent him a message to discuss the locations. He replied that he was traveling and that someone on the ground in Cairo was handling the matter; he would speak to that person and get back to me.

I wrote to Maher that, after consulting with trustworthy members on our page, I thought the Shubra Roundabout would be an appropriate place for protesters to gather and begin the march. Small demonstrations would start in different areas of Shubra and then meet at the roundabout. Days before, Shubra had seen large demonstrations by Christians and the security forces had been incapable of controlling huge crowds there. The Muslim Brotherhood also had a significant presence in Shubra, a residential neighborhood with narrow side streets that made it easy to mobilize people and rally them toward central points. I suggested that we also maintain a backup plan for the activists to head to Imbaba if the demonstration in Shubra was contained. Maher was not convinced and asked me to wait until he coordinated with the people in the field. I knew my field experience was limited, so I was happy to defer to him.

The activists met several times, and Maher sent me summaries of their discussions. He said they were considering starting from the popular neighborhoods, then bringing the crowds toward Tahrir Square or the Ministry of Interior. He also suggested that we limit

Jan25 to Cairo and not include other cities. He said the activists would divide themselves among the neighborhoods in order to act as catalysts to incite participation. He stressed the importance of rallying for Jan25 after the Friday prayer on January 21 and promoting Jan25 on the streets as much as possible.

As that Friday approached, I received messages that emphasized how crucial it was to hold the demonstrations in multiple locations. On January 20, I sent Maher a message saying that we were lost. Nobody was in charge; everyone had his or her own idea about where we should meet. He said that the bright side of being lost is that the security forces are lost too. In retrospect, he couldn't have been more right. Mostafa al-Nagar sent me a message saying that he and his team would organize a march of university professors and other professionals that would start from Cairo University and move toward Tahrir. The idea was very appealing—lawyers, medical doctors, and professors starting from the university would deliver a strong message.

I had earlier written an article with the title "I Wish." I reposted it on the page several times. I wanted to reiterate our goals and dreams in order to keep the big picture burning brightly.

What is it you people want?

Ever since I launched this page, I have frequently received this question in different forms and from different people . . . Some of them were people who only want to frustrate young Egyptians, others were convinced that the country needed to change but did not understand what our goal was, and others did not even believe that the country needed to change and thought that "stability" was very important, and that it would be hard to change Egypt without simultaneously inviting a host of problems that we could never endure.

I thought about this question: "What is it I wish for as an Egyptian?" And I wanted to ask every one of you what you wish for, because I'm sure our problems are different even if their causes are one . . .

Personally, I wish I had a real voice in my own country . . . I would choose my parliamentary representative and my head of state . . . I am tired of feeling that my vote does not matter and that rigging takes place regardless of my vote and opinion . . . I don't want to go to a ballot center and be told by a laughing thug, "It's okay—we have already voted on your behalf." . . . I wish for true democracy, not a sham of a democracy.

I wish we could stand up against corruption . . . I don't want to know that the American government punished Mercedes because that company bribed government officials in several countries, including Egypt, and that my government agreed with the U.S. that the Egyptian official's name would not be revealed to anyone . . . Why is that? Is this not the people's money? When someone steals from your home, is it not your right to find out who he is, and to see him return what he has stolen and receive proper punishment for his crime?

I wish teachers would establish in the hearts and minds of students a genuine love for knowledge and learning . . . not the art of memorizing information . . . I wish our minister of education would be an accomplished professional in the field of education . . . and that the government would support him or her with a large budget, because educating new generations is our only hope if we want to develop this country . . . We have beautiful minds that we waste every year because of negligence.

I wish police officers would be the way they were depicted long ago in the movies . . . respectable men who keep a clear conscience and who are honest before God in using their authority . . . modest yet firm . . . They do not insult poor and helpless people while respectfully addressing the son of a famous businessman . . . I wish they would do away with the secret torture cells at every police station . . . I wish the prosecutors would really represent the people and seriously monitor police officers.

I wish we would stop pulling strings and evading the law for the sake of granting favors . . . I do not want to learn that the son of the head of the Lawyers' Syndicate was appointed a district attorney even though his academic grade was a bare "satisfactory" . . . Neither do I want to learn that so-and-so bought a land plot at a ridiculously low price just because he is the minister's friend.

I wish bribes would become a crime that we would all hate . . . And I wish the person who demands a bribe would be the one to stand out in society, not the other way around . . .

I wish the government would reevaluate its methods of hiring . . . Millions of government employees are rendered idle, and their skills have rusted with time. All they can do now is read the papers and chew pumpkin seeds, if they go to work to begin with . . . I wish that people who work hard would not receive a mere 500 pounds at the end—half of which gets spent on transportation—forcing them to accept bribes and paving the way for unlawfulness to become a way of life.

I wish the government would stop treating people as though they were children who could be lied to . . . As a result, the people have lost all trust in the government . . . to the extent that sometimes even genuine good news is met with disbelief and conspiracy accusations . . .

I wish people would treat one another without classism . . . and that people would quit excessively glorifying other people . . . It's true that we were created rich and poor, but that does not mean we cannot treat one another equally . . . It would be nice if we could learn from people of other nations where you would find the person driving a Mercedes equal to the person driving the Fiat . . . No one has greater priority or respect when it comes to human rights . . . Arrogance afflicts people when they feel those around them are giving them more value than they deserve . . .

I wish government officials would understand that the majority of revenues the state collects in its annual budget is tax money that comes from the people . . . And that a government employee's salary comes from the people . . . So the money you and I pay every month to the company (meaning the government) is what provides his salary . . . He would then start dealing with people as an employee doing a public service, and not as a manager over them . . . Did you ever call Technical Support at a cellular phone company and find the employee treating you as an inferior?

I wish we could rid ourselves of the negativity and passivity . . . I wonder if they planted this in us deliberately? And if so, how they did it . . . We've become a nation that trails behind in almost everything we do . . . We almost only lead other nations in unfavorable things . . . But why? Why do we all say, "This country is theirs. Let it burn, why would I care? I can't fix the world."

I wish we could learn to differ in opinion without insulting one another . . . I wish Muslims and Christians could treat one another with respect and assume the best intentions about one another . . . And not fear the ghost of sectarian strife that we have been fooled into thinking exists . . . I wish we could all realize that we are all Egyptians and that we have equal rights in the world and that it is only God who will judge us in the afterlife.

I wish we could love one another . . . really love one another . . . I wish we would abandon all feelings of unfounded suspicion . . . It simply cannot be true that whenever someone strives to do something worthwhile for their country, they must be spies or funded by foreign powers . . . or, at best, are chasing after stardom . . . Why have emotions of bitterness, hatred, and assuming the worst in people taken precedence over emotions of solidarity, faithfulness, and assuming the best?

By the way . . . It is my right to dream and to pursue my dream . . . And it is your right to dream . . . And, seriously, if we stopped dreaming we would die.

1,689 Likes 1,388 Comments 440,064 Views

As soon as Jan25 was announced on our page, the rate of new members joining each day leaped from five to six hundred to more than three thousand. I posted a message about the importance of uniting forces and acting as one to achieve our dream. The average number of daily comments on the page reached 15,000. The page was a buzzing beehive.

The worry that many shared was that all this momentum was only an Internet phenomenon. One artist expressed this notion in a painting that said, "This is a street, and this is Facebook and not a street." She was implying that the revolution had to be street-bound, and that a revolution on Facebook did not matter.

See you on Jan25 . . . It will be the day we begin organizing ourselves and raising our voices against our condition, on the ground, among the Egyptians, and not on Facebook . . . Jan25 is a beginning that must witness our unity . . . A beginning of a new era of proactivity and chasing after our rights . . . A beginning of a dream, that we each have a voice and an opinion . . . Let us not inflate our optimism and expectations . . . Not everything we seek will happen on Jan25 . . . It will certainly not be the last day of injustice and corruption . . . But if we join forces and act as one, it will be the beginning of the end.

393 Likes 135 Comments 107,699 Views

Meanwhile, the NDP's Electronic Committee launched a campaign to dissuade people from taking to the streets. Its members claimed that the people who planned to participate were thugs and vandals. Our emphasis on the peaceful manner of our protests was crucial in countering this defamation campaign.

Important for everyone who will participate on Jan25: We are not promoting chaos or destruction or attacks on any public or private property . . . We are taking to the streets to demand our rights; we will protest, stage a sit-in, and defend ourselves only if we are attacked . . . This is very important. Let us all please spread this everywhere online.

674 Likes 163 Comments 137,520 Views

One of our members wrote a moving comment in which he stressed how critical it was for Jan25 to succeed. I published it in order to motivate people and counter all defeatist thoughts.

If Jan25 were upon us and no one did anything, and this whole Internet campaign came down to nothing, then the Egyptian youth will look like crap. The world will look down upon us and so will our own people. I hope Jan25 doesn't turn into a big scandal and we only find 100 people on the street who will be rounded up in minutes by Central Security and that would be the end.

715 Likes 538 Comments 125,245 Views

On January 21, the press carried the story of the death of one of the Egyptians who had set themselves on fire following the example of Bouazizi of Tunisia. Some of the activists on Twitter and Facebook started demanding that we not wait for Jan25 and that we take to the streets "now." A few did head to Tahrir Square, but the security forces were prepared. Some skirmishes took place, and that was the end of it. A young woman who was there, Asmaa Mahfouz, showed rare courage. When she found only a few others at the protest on that day, she decided to make a video to motivate everyone to participate on Jan25, which we put up on our page right away.

The video was very effective in encouraging many page members to break their own fear barriers. They were impressed by Asmaa Mahfouz's brave mention of her name as she demanded that everyone take action against injustice, corruption, and oppression. Most comments on the video were positive and promising. Many members described Asmaa as a "girl worth a hundred men or women," which is a common Egyptian expression to denote courage.

Communication and coordination continued. I corresponded with a group founder on Facebook, Ibrahim al-Masry, whose group included more than one thousand amateur photographers. I was a big fan of his work and that of his amateur friends. He now expressed a strong interest in helping. On January 19 I anonymously asked him to get involved with the protests. He was immediately responsive and created an event that he called "The Camera Is My Weapon." More than a hundred photographers joined Jan25 as a result, in order to take pictures of the action. This was an important way to defend the protesters: the regime feared the media more than it cared about the welfare of protesters, so the security forces would be careful not to be photographed showing any violence.

News of suicides on the streets began to intensify. I Googled the suicide phenomenon in Egypt and shared a devastating report I found:

> Accomplishments of the Egyptian government: 12,000 Egyptians committed suicide during the past four years! 5,000 committed suicide in 2009, and more than 100,000 suicide attempts occurred during the same year! This rate is five times the rate of 2005! Every day 14 Egyptians commit suicide, and the reasons are unemployment and poverty, and more than two-thirds of the cases are young men under 25 years of age. And the source of this information is the Central Agency for Public Mobilization and Statistics.
>
> 278 Likes 127 Comments 114,521 Views

I was adamantly against suicide. Yet I was ready to die for the cause if that was the only way to change my country. Suddenly, and with little deliberation, many Egyptians had become entirely dedicated to freeing our country from the tyranny of an authoritarian regime.

> I feel exceptionally optimistic and hopeful . . . For the first time, we are hearing Egyptians on the street discussing Jan25 and asking one another if they will participate . . . For the first time, 60,000 Facebook users confirm attendance . . . For the first time, an invitation reached almost one million Facebook users . . . For the first time, all this momentum is created . . . For the first time, I feel that I am prepared to die if this will bring back the rights of my countrymen.
>
> 711 Likes 170 Comments 116,696 Views

On Thursday morning, January 20, I posted a notice that the protest locations would be announced at midnight, in order to make sure that the congregations at Friday prayers—the following day—could promote Jan25. Friday prayers were by far the largest weekly gatherings of Egyptians from all classes.

Some of the active political opposition forces, who had formed a "popular" people's assembly to counter the fraudulent 2010 Parliament, had announced their support for Jan25. They declared that their march would start from in front of the High Courts Complex. The media had played up this announcement, although the regime ignored it, just as it ignored our Facebook announcements.

> Members of the Popular Parliament decided to organize a march on Jan25 outside High Courts Complex with a list of demands . . . The Popular Parliament was founded by a group of different political opposition forces and independent personalities after the rigging of parliamentary elections . . . Thank God, many people are taking action.
>
> 310 Likes 57 Comments 98,139 Views

I was very happy to learn of their march, and it didn't bother me that they did not wait to coordinate with anyone. Looking back, the glory of the Jan25 revolution was that the invitation did not originate from any political body or organization. There was no plan that rendered everything else as deviant. In harmony with that spirit, I used "*Kullena Khaled Said*" to support anyone who saw that a different form of action was needed to express disapproval of the regime.

> Other groups and Facebook pages have announced locations for gathering and marching other than the ones we agreed upon . . . This is great . . . Because there are so many people who want to participate . . . And to be scattered will serve two purposes . . . First: All Egyptians will find out about Jan25 and participate . . . Second: Security forces will not know what areas we will cover, or they will disperse their forces and not concentrate their power in one area, allowing them to control the demonstrations . . . Thank you to all the proactive youth.
>
> 494 Likes 120 Comments 131,805 Views

Our problem with the police was not a personal one. I did not want the demonstrations to turn into physical confrontations. This is why I posted, several times, news of good, honest police officers. I published the name of an officer who refused to accept a $15 million bribe, and we celebrated him on the page. I also apologized for accusing one officer of failing to protect the Alexandria Coptic church at the time of the bombing, because I discovered that he had been severely injured. I saluted him and extended my respect to every honest police officer.

Just as we fight corruption inside the police force . . . we must support the good . . . The police force includes many officers who died honorably doing their job, which is to protect you and me . . . We do not oppose the police force as a whole, we only oppose their violations of human rights . . . I will be the first to take pride in a brother who is a decent police officer . . . We must support the good just as we fight the evil.

605 Likes 125 Comments 108,950 Views

Time was running out. All of the activists had to agree on the locations that were soon to be announced. Mahmoud Samy gave me the good news that the activists had indeed all met and coordinated everything: the protests would start in Arab League Street in Mohandeseen. Mostafa al-Nagar also confirmed that professional doctors and lawyers would participate in the Cairo University protest, but he said that the location might change at the last minute. Ahmed Maher agreed to convene in Shubra and then to march toward Tahrir. As for me, I chose Matariyya Square as a protest location for the other half of Cairo. I announced all four locations on the page. The information started to spread.

Our page members began to take photographs of themselves to confirm their participation on Jan25. Many designers also e-mailed in great designs for Jan25.

> Now, two crucial factors will determine whether we succeed or not. First, we must all vow to participate and not count on the sufficient participation of others, because this is often the reason for limited turnout . . . Second, we must deliver the message to everyone in popular neighborhoods so that even if they do not join us, they will protect us . . . We must all move on the ground now and not talk to one another over Facebook.
>
> 348 Likes 68 Comments 107,920 Views

Rumors started to spread. Some controversial figures who maintained an online presence began to circulate unfounded speculations that Gamal and Alaa Mubarak, the president's sons, had escaped the country along with several businessmen. Someone sent me a message asking that I publish these rumors for motivation, but I emphatically refused, and instead decided to lead a campaign against spreading false news, even if it served our cause.

> I'm receiving many suggestions to publish rumors that serve Jan25 . . . My answer is that I will never publish a lie as the truth because I will never in my life adopt the principle of "the end justifies the means" . . . You can call me a lousy politician or a naive person, but I have principles, and one of them is to be respectable and keep a clear conscience. I fear God and don't care whether I lose or win.
>
> 600 Likes 164 Comments 123,933 Views

Khaled Said's mother pledged her support for Jan25. One of the activists from Dr. ElBaradei's campaign filmed a short video in which she urged everyone to take to the streets on that day and participate in the peaceful demonstrations. Expatriate Egyptians in Europe and the United States were also supportive. Many of them sent pictures of themselves holding signs that revealed their advocacy of Jan25.

At his army camp, AbdelRahman Mansour was not aware of how

quickly things were developing until his brother and a friend visited him on Friday, January 21. Both his visitors knew that he was an admin on the page. They wanted to give him the good news: it looked like the event was now going to have over 100,000 confirmed attendees and the invitation reached over a million Egyptian Facebook users before Jan25. AbdelRahman was elated, but he regretted being unable to join the protests on that coming Tuesday.

Nothing I could ever write on paper can accurately convey what it felt like to witness the buildup of such immense solidarity. Egyptians were ready for something big. Not all of us knew it at the time, but after many years of onerous opposition, we were finally ripe for revolution.

6

January 25, 2011

MY FLIGHT TO CAIRO was set to leave January 23 at 4 A.M. Hours before my departure, my anxiety level began to rise. As a contingency plan, I sent a message to Nadine in the United States telling her that I was leaving for Cairo and that if something were to happen to me, she should wait a week before announcing my relationship to *"Kullena Khaled Said."* We had agreed that Ahmed Saleh, an activist whom she knew well and who was a friend of both AbdelRahman Mansour and Mostafa al-Nagar, would update the page during my absence. I wanted to ensure that the page would be constantly updated if I was arrested, and I didn't want anyone to suspect that I was the admin. I also called Najeeb, one of my very few Google coworkers who knew about my online activism, and provided him with access to the page's admin account, so that he could update the page if he noticed that I hadn't updated it for more than twelve hours. A week before, I had asked my manager, who is based in Turkey, for a short vacation so I could go to Cairo for personal reasons. *If turnout at the demonstrations is small, then I shall return after two or three days,* I told myself. *But if the protests are large, then my short vacation and few clothes aren't going to get in the way of the liberation of Egypt!*

I spent some time with my kids before they went to bed, and I explained to Isra that the purpose of my trip was to attend a protest against Hosni Mubarak. I told her that Egyptians no longer wanted him as their president. Isra knew who Mubarak was from the many conversations I had had with her about the situation in Egypt. I explained to her that Mubarak and his government were directly responsible for the country's miserable condition. Adam, however, was too young to fully grasp the reasons behind Jan25. I asked them both to treat each other nicely and not to give Ilka a hard time during my absence. I then hugged them and told them that I was going to miss them very much and that I would be back as soon as I could. Of course, I didn't really know when I would be back.

As I was getting ready to leave, I saw the apprehension on Ilka's face while she watched me pack. She was not only worried; she was also very disappointed. I could fully understand how she felt. To her, everything I had done in the past few months had come at the expense of our personal life and our children. She believed I had been selfish because I had not thought of the consequences of my actions for her, Adam, and Isra. She asked who would take care of them if something were to happen to me. I said she was right and I understood how she felt, but it was too late, as I was already in too deep. I could not invite others to demonstrate and then sit at home and follow the news from afar. I tried to assure her that I would be safe.

Before leaving, I went back upstairs to see the children. They were already fast asleep, so I kissed each of them a final time. Ilka went into our bedroom, and as soon as I came out of the kids' room she handed me a green wristband that we had picked up at a charity event for cancer victims. She asked me to wear it at all times while protesting, to think of her and the children every time I saw it, and to stay safe for the family. I held her tightly and bid her farewell with the promise that I would do my best to remain unharmed.

I will never forget my trip to the airport that night. I felt strength mixed with fear of the unknown. I played and replayed Hans

Zimmer's soundtrack from the movie *Inception.* The experience was emotional, with the car cutting through the night with suspenseful music playing in the background. For that half hour, memories of my political involvement flashed through my head, starting with my support for Dr. ElBaradei and going on to creating *"Kullena Khaled Said"* and organizing the Silent Stands and protests, and then ending with Jan25, which was now forty-eight hours away. I began to murmur prayers and prepare myself to die for the cause.

At the airport, waiting at the gate for boarding, I went online to catch up on the page's activity. I logged into the page's e-mail account and started to delete all the sensitive chat conversations related to Jan25 preparations; I didn't want to give the security forces in Egypt a chance to know any details about the conversations I had had with the key people working behind the scenes. I had insisted on posting a message every half hour at least, so it was disturbing to think that I would have to spend four hours—the duration of my flight from Dubai to Cairo, including customs and immigration—without updating the page. But I decided to force myself to sleep during the flight time to make the best of my time offline.

After sleeping during most of the flight, I wrote a brief message to Najeeb before I reached passport control. It said, "I am now being arrested at Cairo Airport." The message was only a draft; if I needed to send it, all I had to do was press the Send button.

The passport control officer smiled at me as he took my passport and asked me to sit while the usual security procedures were completed. I had become accustomed to this treatment, since my name was still on the airport arrivals watch list. Ten minutes later, a security guard escorted me to the Red Hall to inspect my luggage. *Nothing new so far,* I thought.

The guard asked me the question I had grown used to: "What have you done, sir? Why is your name on the watch list?"

"I don't like Hosni Mubarak, and that's why my name is on the list," I said. He murmured his agreement; he did not like our president either. "God willing, this will be the last time I'm searched, because we will catch up with Tunisia on January twenty-fifth," I said confidently.

He gave me a diplomatic smile and carried on with the routine search without any further discussion.

I left the airport and headed to the Radisson Hotel. I had decided to stay there, as I usually did when my Cairo business trips required focus, without the distractions of staying at my family's house. There was also a security reason this time. If the police went to my family's place to try to arrest me, it would be possible for me to receive a warning and prepare myself, since I was staying somewhere far from home.

I continued to rally supporters on the page. I shared writings, images, and designs made by the page members that promised a victorious Jan25. I tried to induce confidence and belief that youth would be capable of leading change. I also highlighted the government's frightened reactions.

I decided to compile all the information relevant to Jan25 in a document that was easy to print and to distribute online. It summarized the reasons for protesting and for choosing this day and these locations. It also described the unified chants that had been chosen and provided phone numbers for activists responsible for supporting arrested protesters and for redirecting demonstrators to other locations if the protests at any one place were obstructed. I uploaded the file to Google Docs, where more than 50,000 people accessed it. Its content was also disseminated through various online forums, political websites, Facebook pages, and Twitter accounts.

Everything You Need to Know about Jan25

Important note: Please visit the page regularly for frequent updates.

Who Are We?

The call for protests on Jan25 began from the Facebook page *"Kullena Khaled Said,"* a page created to support the case of Khaled Said, who was beaten and tortured to death on the street in Alexandria in June of 2010. The call was spontaneous and not planned by any political or popular force. After the invitation spread, and because of events in Tunisia, Many Egyptians were encouraged to participate and spread the word. The Facebook page is not influenced by any political party, group, movement, or organization. It is independent and does not support any person or specific ideology. It belongs to all Egyptians who wish to defend their rights. The page is managed through its member volunteers, and this is the secret to its success.

Why Protest?

Egypt is going through one of the worst periods in its history on all fronts. Despite the reports propagated by the government that aim to polish its image, reality is unfortunately different. Our collective participation on Jan25 is the beginning of the end—the end of silence, acceptance, and submission to all that is happening in our country, and the beginning of a new page of coming forward and demanding our rights. Jan25 is not a revolution in the sense of a coup, but rather a revolution against our government to let them know that we have taken interest in one another's problems and that we shall reclaim all our rights and will not be silent anymore.

There are 30 million Egyptians suffering from depression, including 1.5 million with severe depression. There were more than 100,000 suicide attempts in 2009, resulting in the deaths of 5,000 people. There are 48 million poor citizens in Egypt, including 2.5 million who live in extreme poverty. 12 million Egyptians have no shelter, including 1.5 million who live in cemeteries. Habitual corruption has led to graft totaling more than 39 billion pounds during one year alone. And Egypt ranks 115th out of 139 nations in the Corruption Perceptions Index. More than three million young Egyptians are unemployed; the unemployment rate among youth has surpassed 30 percent. Egypt also came in last place among 139 countries in terms of hiring transparency. We have the highest rate of newborn deaths in the world: 50 of every 1,000 newborn Egyptians die. Half of Egypt's children are anemic, and eight million Egyptians are infected with hepatitis C. More than 100,000 citizens suffer from cancer every year because of water pollution. Our rate of ambulance cars to citizens is 1:35,000. Egypt's emergency law has caused the death of many Egyptians as the result of torture and also facilitated the arrests of thousands without official warrants. Security forces' involvement in monitoring politicians and blocking their activism has led to shameless rigging of parliamentary elections, producing a people's assembly that was more than 90 percent controlled by the ruling party . . .

Why Jan25?

In 1952, using simple rifles, our forefathers in the police force fought the British army's tanks. Fifty became martyrs, and more than 100 were taken as prisoners. They were heroic and offered a model of sacrificing one's life for one's homeland. Now, more than 50 years later, we suffer from a police force that has become a machine for torturing and humiliating Egyptians. We chose this day specifically because it represents the fusion of the police and the people. We hope that decent and honest officers will join our ranks on Jan25 because our cause is one with theirs. Jan25 is a national holiday, which should allow all Egyptians to participate without hurting their productivity on any level.

What Are Our Demands?

Demand One: Addressing the poverty problem before it explodes by respecting the judiciary's verdict to increase the minimum wage fairly, particularly in health care and education, in order to improve the services offered to the people. Also, to issue unemployment benefits that reach up to 500 Egyptian pounds, for a limited period, for every university graduate unable to find a job.

Demand Two: Annulling the emergency law, which has led to the control of Egypt by the security apparatus and has allowed the illegal arrest of political opposition members without cause. We demand that attorneys general take charge of police stations to stop the routine torture practices that take place there. We also demand implementation of the judiciary's verdicts by the government.

Demand Three: Firing Minister of Interior Habib el-Adly. His loose security measures have led to terrorist attacks as well as the spread of crime at the hands of police officers and other Ministry of Interior forces.

Demand Four: Placing a two-term limit on the presidency.

Absolute power corrupts, and no advanced nation allows a president to stay in power for decades. It is our right to choose our president, and it is our right not to have a president monopolize power until his death.

Of course Egyptians have many further demands in areas such as health and education. We begin by moving together, pressuring the government, and achieving one demand at a time. This is our role as a people: to direct the government, hold it accountable, and determine its priorities.

Time and Place of Protests

It is very important to realize that the point of the protests is to mobilize all people to our side. Everyone is frustrated and dissatisfied with the nation's condition. We must encourage them to participate. This is why we must organize marches in all working-class neighborhoods. And people must move together in numbers of more than ten until they reach the protest destination. And by the way, the locations are not just the ones mentioned here. There are other locations that have not been announced where protests will be organized in different governorates. The important thing is for you to come out and express your anger in whatever way you can.
Greater Cairo: Shubra Roundabout, Matariyya Square, Cairo University, the Arab League Street
Alexandria: Al-Manshiyya Square, Mahatet Masr Square
Important note: Other groups are organizing protests and marches in popular neighborhoods all over Cairo, Giza, Alexandria, and many other cities across Egypt. If you are in a working-class neighborhood, come out on Jan25 and join them . . .

Protesting Guidelines

1. The protests are peaceful. We are peace advocates and not advocates of violence. We are demanding our rights and must uphold the rights of others. We will not respond to any provocation from security forces and lose control. This is what they want us to do. One of the security forces' main goals is to portray the protesters as thugs who want to destroy our country. We must discipline ourselves and refrain

from foolishness or any violations of the law, and we must not endanger any person's life or cause harm to any public or private property. If you see someone behaving violently, please circle around this person and take him out of the protest.

2. Please be at the protest location promptly at the determined time. Delays will hinder our efforts and risk the failure of the protest. Being present at the same time makes it easy to start the protest and makes it hard for the security forces to prevent it.
3. When leaving your house, do not carry anything you don't need, such as membership cards or licenses or credit cards. Carry only your personal ID and a sum of money sufficient for an emergency. Please don't bring your watch or anything easily breakable. The ideal attire would be sportswear or jeans along with a jacket to keep warm in case the protest runs long or develops into a sit-in. Please bring a large bottle of water, because there is always a shortage of water inside the protest.
4. Please carry the Egyptian flag and refrain from carrying any signs of a political party, movement, group, organization, or religious sect. Jan25 is for all Egyptians. We are all demanding equal rights and social justice and do not want to be divisive.
5. If you are not an experienced protester, leave the front lines for experienced protesters to lead the march in order to avoid conflicting decisions.
6. The chants are unified and agreed upon. Please refrain from all profanity and do not enter into quarrels with security force members. Central Security is not your enemy. They are guards who have been forced to spend their compulsory military service in this capacity, and if they disobeyed the orders they would be punished badly. Try as much as possible to target your anger at the real enemy.
7. Try as much as possible not to disturb traffic. We are not out to punish the citizens but to demand our rights. Try not to interrupt traffic flow deliberately. It is understandable that when tens of thousands of people take to the streets, traffic flow will be affected. This is not what we are referring to.
8. Do not come out alone. It is very important not to come out

alone, because friends come in handy in situations like these. Please be with someone and talk someone into coming out together, just like we take to the stadiums when there is a match to watch.

Unified Chants

The unified chants are one of the most important protest ideas. We are all out for the sake of Egypt, and we must unite and act as one. We will all stick to the chants together and will focus on the issues of unemployment and poverty. These are the issues that concern all Egyptians. These are the chants that have been agreed upon:

- Long live Egypt . . . Long live Egypt
- Bread . . . Freedom . . . Human dignity
- Dear freedom, where are you? . . . Emergency law is keeping us away from you
- We will not fear, we will not bow . . . Like we've done for so long
- Beloved people of Tunisia . . . The revolution sun will not set
- We shall sacrifice our blood and lives for you, dear Egypt
- Raise your voice up high . . . With injustice we will not comply
- Alert the people and rock the universe . . . We will not let our homeland go
- Minimum wage now . . . before all the Egyptians revolt
- It's my right to find work and live . . . The petty income is not enough
- Let's make it happen, Egyptians, wake your spirit . . . The gates of liberty are open
- Let's go, people, transcend the fear . . . Let the whole world know
- A people with the civilization and glory of years . . . Will not bow until Judgment Day

1,977 Likes 1,527 Comments 506,871 Views

On January 20 I had sent Amr El-Qazzaz, one of the cofounders of Rasd, the brilliant online parliamentary-elections monitoring service, a message asking if he thought the Muslim Brotherhood would officially participate on Jan25. If its members took part, then their enormous numbers would certainly help us immensely. He knew for certain that the MB would not participate officially but said that some of its young members and a few key leaders might participate as individuals. I asked if there was someone I could contact anonymously, but he did not think it was possible—the MB would not make plans with an anonymous person whose objectives they did not know. I understood and respected his opinion. I gave up the idea of coordinating with the MB. I still hoped that the organization's younger members would come out on Jan25, especially when I found some of my Facebook friends who were members of the Muslim Brotherhood promoting Jan25 and announcing their participation.

Amr and I also had a lengthy discussion about Rasd and its method of operation. I criticized its noticeable bias toward the Muslim Brotherhood and told him that what I liked most about the name *Rasd* was its neutrality. He asked me to promote the Rasd page before Jan25, because he and the others planned to report on the events intensively. This is when I thought of recommending that they create a new page, for two reasons. On a practical level, the current title was "2010 Parliament," and Facebook did not allow title changes. But more important, a new name would represent a new beginning without any of the old impressions of Rasd.

I had earned Amr's respect because of the success of "*Kullena Khaled Said*," and he seemed convinced by the idea I proposed. I explained that Rasd's neutral role was different from our Facebook page's mobilizing role. Rasd had to be a source of information and not a source of analysis or bias. "If Jan25 fails and there are only four people on the street," I explained, "your role as Rasd would be just to say, 'Four people protested and here is a picture,' while '*Kullena Khaled Said*' would rally for the people to come out on the twenty-sixth." We also discussed documenting the news with images and videos. During the elections, Rasd had published extensive news that it was not able to document. I told him that the page, which was an unknown source to most people, would become credible only if it showed proof together with the news reports. On Jan25, Rasd must not publish information unless its members had supporting evidence to publish with it, I said.

Shortly after this conversation, Amr sent me the link to Rasd's new page. I said I would post it on our Facebook page throughout Jan25 and recommend that people use the page to follow the news updates. Similar collaborations were taking place among many activists in diverse locations, in a collective attempt to make Jan25 as big a protest as possible.

As Jan25 approached, members' contributions increased. Scores of photographs expressing their senders' intentions to protest on the big day reached the page. The most touching of these images was one of two young men, a Muslim and a Christian, and the caption said:

> I am Amir Botros, a Christian Egyptian who will protest on Jan25 to demand justice for Sayyid Bilal (a young Muslim man tortured to death by State Security, after the Jan 1st church bombing) . . . And I am Youssef Ahmed, a Muslim Egyptian who will protest on Jan25 to demand justice for Maryam Fekry (a young Christian woman killed in the bombing)
>
> 1,936 Likes 429 Comments 124,868 Views

This kind of daring attitude compelled many others to take pictures and send them. On January 24 the Internet was flooded with images of Egyptian youth who had decided to protest. Many people declared that they were ready to die as martyrs for the cause. The most important task of January 24 was to stress the next day's guidelines: that the protest was to be peaceful, chants were to be unified, and people were to refrain from promoting political parties or groups. I continued to rally and motivate as many people as I could to participate.

> Every kind of logic, rationale, and religious or ethical consideration says we must all take to the streets now that Egypt is going from bad to worse . . . The rich are getting richer and the poor are getting poorer . . . We suffer from polluted air, cancer-causing food, and contaminated water . . . unemployment, lost youth, drugs, and increasing suicide . . . corruption and embezzlement of public funds . . . Egypt has become one of the least developed nations on all fronts . . . Family members are losing their lives to cancer every day . . . Come out and say, Enough, enough, enough!
>
> 529 Likes 266 Comments 143,959 Views

My rallying was not limited to sentences I wrote and uploaded. One of the members sent me an old speech by Mubarak delivered on a religious occasion in the 1990s. Out of context, it sounded as if Mubarak was motivating people to participate on Jan25. I posted it and commented that President Mubarak was inviting all Egyptians to fight passivity and take to the streets on Jan25. In a matter of a few hours, the video had spread to thousands of users.

Here is an excerpt from the speech:

> How brilliant is this saying by the honorable Prophet: "If you witness a transgressor and refrain from setting him right, God's infliction of overwhelming suffering upon you will be imminent." This is such an expressive image that the Prophet, God's peace and blessing be upon him, illustrated to depict the evil of transgressing against someone's rights under the pretense that to do so is the transgressor's right.
>
> The Prophet then reveals the consequence of passivity in addressing transgression against the right of a community. It is a grievous and severe consequence because it leads to the demise of the transgressor and all the silent, passive community members. In the image, the community is represented by a people aboard a boat, where they have divided the space among them. Then one passenger began to wreck the space underneath him under the pretense that it was his alone.

The honorable Prophet urges us to face this threat against the peace and security of the community: "If they set him right, he survives and they survive; if they neglect him, he shall perish and so shall they."

678 Likes 171 Comments 101,565 Views

Obviously Mubarak supporters didn't agree with that. The NDP's Electronic Committee intensified its activities noticeably halfway through the day on January 24. New accounts were created and made to seem as if they belonged to opposition parties and antiregime political movements. The false accounts all attempted to deflate people's enthusiasm and described Jan25 as a conspiracy leading the nation to its destruction.

Discussions of Jan25 spread to every street and media source. Loyalist columnists denounced Jan25 as a scheme to defame National Police Day and described opposition leaders as disrespectful of the Egyptian police's honorable history. Nevertheless, they insisted that Egyptians retained the right to demonstrate. All of this supported our promotion campaigns for Jan25 and served our purposes well.

Also critical to mobilization efforts were the writings of public personalities who were not famous for political activism. Many of these were singers and actors who decided to participate on Jan25 and to promote the event. Among them were the singer Hamza Namira, the actor Amr Waked and the actress Basma, the TV presenter Ahmed el-Esseily, and directors Khaled Youssef, Amr Salama, and Mohamed Diab.

It was clear that Jan25 had caught the attention of various international media organizations. I received an e-mail on January 18 from a *Newsweek* reporter named Michael Giglio. "I'm interested in highlighting the work you've done with your page as a way of discussing the situation in Egypt and how it has been affected by the events in Tunisia. I'd also like to bring attention to the demonstration that's being planned for Jan. 25," he wrote.

"Unfortunately, I don't do phone interviews," I responded. "My identity needs to be kept confidential, even with journalists. Our page is tightly monitored by security police in Egypt. If you want, I'm happy to do an e-mail interview. Otherwise, I'm better off losing a story that might reveal my real identity." We met halfway and agreed to converse via Google Chat. After about an hour of discussion, I sent Michael a lot of pictures and stats from the page. A few days later he sent me excerpts from his article, "Is Egypt Next?," prior to its publication. He wanted to make sure I felt his depiction was accurate and did not compromise my efforts in any way. It was a positive and well-balanced story, and it ended with my answer to his question about what I would do if Jan25 failed: "We're hoping a lot of protesters turn up, and that people in the streets see us, connect with our demands, and join us." And if the effort failed, "I'd learn from the lesson, move forward, and do something else."

On the night of January 24, I agreed to grant Amr El-Qazzaz of Rasd the right to update *"Kullena Khaled Said"* on Jan25 under the condition that he would do so only for urgent reasons: to warn against certain dangers on the ground, for example, or post news that would serve as a strong motivation for the protesters. He would do so in parallel with updating the Rasd page, together with his partners.

At the break of dawn on Jan25, I placed an advertisement for Rasd's new page on our Facebook page. Rasd's new members numbered only a few thousand at that point. Then I wrote a final message before going to bed, which included a link to Hamza Namira's song "Dream with Me." It was a call to everyone to dream of a better tomorrow that we would share in making.

I also posted a farewell message that night to all the page members, and I decided to repost it in the morning. I had no idea what was going to become of me or anyone else. So I decided to write something that would serve as a memory of me after the revolution had ended.

I have no idea what will happen today . . . I have no idea where I might end up tomorrow night . . . at home . . . striking somewhere on the street . . . arrested in a prison cell . . . buried at a cemetery . . . But what I do know is that I must reclaim my rights and my children's rights from every person who has humiliated Egyptians in this country . . . Because our country is not a piece of property that can be divided among a few thousand men while we sit and watch . . . Because I sleep on a full stomach but I know well that 30 million Egyptians go to sleep dreaming of something to eat . . .

I am going out on Jan25 because this is my country and because I am no longer going to accept the fact that I'm so insignificant that my fate can be decided by a government, or State Security, or a police force . . . I am an Egyptian citizen . . . You know Egypt? The Egypt that defeated Israel in '73 . . . The Egypt that challenged Britain . . . and expelled the French . . . The Egypt that stood strong in the face of all occupiers and none were able to abolish its identity . . . I am going out on Jan25 because I do not agree that Egyptians are cowards . . . The Egyptian people are not cowards . . . Egyptians are worried that if they take action they will be acting alone, without anyone's help . . . If we need to sacrifice, I swear I am prepared to sacrifice everything that I own for the sake of my country.

There are two reasons for concealing my identity since the launch of this Facebook page: a strictly security-related reason, which was to prevent security forces from closing the page . . . and the second, more important reason was that I have no ulterior motives . . . I want nothing from all of you . . . To everyone who is skeptical of my intentions: I swear by God that I have no other motive than for my country to change . . . I want nothing from you, not gratitude and not praise . . . I want no public position or reward, and even if we are successful, I will never reveal my identity, because I wish for us all to carry this beautiful dream into our lives . . . A story with a happy ending . . . My only reward is that someone who does not know me . . . and whom I have never seen . . . will pray for me and say: God bless you . . . This is my only goal.

I sometimes have caused harm to members of this page without meaning to . . . Please, do not hold grudges against me, because I swear that my intention is pure and that I did everything in accordance with my conscience . . . Believe me, I have not posted here any news or image or video unless I was completely convinced it was truthful . . . I never wrote anything to deceive you . . . I wrote here with my heart, not my keyboard

. . . Every word written here came from a watching conscience, a thinking mind, and a judging heart.

You are all my brethren . . . Even those who see me as a spy and traitor funded by enemies of Egypt . . . Because they have been brainwashed, unfortunately . . . They are not my enemies . . . My real enemies are those who took control of this nation and considered it their personal property for 30 years . . . They robbed every square inch of it and considered anything they did a favor to the people . . . The very people who supposedly control the decisions . . . They exploited poverty, hunger, and lack of education in recruiting the youth to their lines and having them attack any voice of opposition . . . But believe me, their end is near.

Please, I ask you all to forgive me if something happens to me tomorrow. And I ask you all to take part tomorrow, because Egypt will not change if we continue to merely chant on Facebook.

2,105 Likes 1,429 Comments 173,861 Views

Before I left the hotel early in the morning, Mostafa al-Nagar told me that the Cairo University protest location had been changed to the Doctors' Syndicate near Tahrir Square. I didn't immediately post the update on the page, however, hoping to delay the arrival of the security forces at the scene. Then I called my brother, Hazem, to ask if he wanted to join me for Jan25. He said yes. Hazem, a dentistry student, is eight years younger than I am. He had never been interested in politics before Jan25. Like many others, he was an example of a young Egyptian who was demonstrating for the first time in his life.

As soon as I arrived at my family's house to pick up my brother, I went online and checked what was happening. I called a friend, who told me that Arab League Street was packed with police forces and Central Security trucks. It would be impossible for a protest to take place there.

Mahmoud Samy, who organized the Arab League Street chapter with his activist partners, was online monitoring the situation as well, and I tried to convince him to change the plan. He told me not to worry. "The participating crowds are going to be huge," he said, "and they will approach the street from the surrounding neighborhoods. It will be impossible for the police to meet such numbers with violence."

Warning comments began to permeate the Facebook page to scare away the members. Regime loyalists were spreading news that thugs and extremists were being arrested and that high counts of injuries and deaths among protesters were expected. This was the regime's final attempt to thwart the demonstrations. They thought that psychological warfare would save them.

I called Amr Salama and asked, "Will you participate?" He said yes, and I asked what he thought was the best location to convene. He suggested we head toward the Doctors' Syndicate. I tried to convince him to go to Arab League Street, but he thought that the protest at the syndicate would probably be much larger. One of my Facebook friends, Yasmina, called me from Arab League Street around noon and said that security forces had covered the place and there were no signs of a protest. At that point I decided to leave with Hazem for the Doctors' Syndicate to meet Amr Salama.

I told Hazem to park the car halfway to our destination, because I thought it would be too risky to take the car downtown. We parked and took a taxi to the Doctors' Syndicate, arriving at the scheduled time for the protest. There were no security forces, but our numbers were very limited—no more than two hundred people. We began chanting and inviting others to join us, but it appeared we were not persuasive enough. Our numbers increased only slowly. There were many familiar faces: Amr Salama was there, along with Mostafa al-Nagar, the poet Abdel Rahman Youssef, one of the leading Muslim Brotherhood figures Dr. Abdul Moneim Abol Fotouh, the satirical columnist Bilal Fadl, and many others. We were mostly using the predetermined chants as agreed.

Suddenly troops of Central Security guards lined up in two rows to block the protesters' way. There were many of them, and I was extremely disappointed. I felt then that the day might end up being like any other, when small protests were easily thwarted. I started to talk to the guards: "I'm here for your sake, for people who work long days to end up with no more than three hundred or five hundred pounds a month. This is simply unfair." Meanwhile they cordoned off the area we occupied, making it difficult for a group to enter or exit. There weren't all that many of us, but everyone was chanting.

Mostafa al-Nagar and others brought flowers to give to the police officers. We chanted, "Long live the police and the people." It was our attempt to let them know they were not our real enemies, and that the real enemy was whoever had brought them to the front lines to face a peaceful demonstration. However, there seemed to be no life on the other side to accept or even hear our cries of outreach.

Despair started to creep in, and I used my phone to write on my personal Twitter account, "Protests are like a watermelon. You cannot anticipate its taste until you open it. I got the bad melon." This was a comical Egyptian metaphor to describe unpleasant surprises. The stalemate continued for more than an hour, and the activists began to discuss forcing our way out of the security cordon. Al-Nagar and some of his companions were trying to negotiate with security officers to let the peaceful march pass through to Tahrir Square as news started to arrive that thousands of people were in the square, which was barely a ten-minute walk away. They offered to walk on the streets without chanting, but their offer was categorically rejected.

Some of the zealous youth in the protest decided to break out of the siege. They sprinted toward the lineup of guards and broke through it head-on. The ring of guards split, and their control over the situation was lost. The young men dashed as fast as they could toward Tahrir Square.

Amr Salama was in the front lines as the protesters pushed past the police. He described his experience in a note on Facebook that circulated widely at the end of Jan25. It was titled: "Why Was I Beaten?"

We were demonstrating in front of the Doctors' Syndicate on Kasr el-Eini Street surrounded by a ring of guards and officers. We wanted to reach the other groups of people at Tahrir Square.

At 2 or 3 p.m. we decided to head toward Tahrir and to take whatever risks were necessary.

Hoping to create momentum and break through, I was one of the people who pushed the guards, and indeed we broke through and dashed toward Tahrir and the Parliament building.

The street was completely empty, and on the horizon I saw huge masses of people. I thought they were protesters until I noticed they all wore black and headed toward us with black sticks. I recalled the war scenes from movies like *Braveheart* and *Gladiator* and I finally knew what it felt like to be in a war from epic times. I found myself spearheading the charge toward them. Others tried to escape from side streets, but it appears they got surrounded and returned. There was a moment of confusion and hesitation, and then the security forces charged toward us.

My dear iPhone was in hand, and I tried to video all of this.

Until a decent number of guards surrounded me and began attacking me with their sticks. They struck at my head, face, stomach, and legs.

Their champion of a leading officer entered the scene, and I shall not forget his face until I die. He began punching my face in a way that I did not think a human body could endure. He took my dear iPhone, may it rest in peace, and crashed it on the ground and stomped on it repeatedly. He then came to his senses and said, "Let him go, stop striking." I thought, *Thank God, his conscience has reawakened.* But he continued, "Because of the cameras." He grabbed me and took me to a side street, where, on our way, we found a young man lying on the ground soaking in rivers of blood from his head. The officer said this, word for word: "Here is one son of a . . . He's dead. I swear I will kill you like him, you son of a . . ." Then he entered a building together with his pleasant guards and locked the gate. He kicked my legs to throw me off balance and I dropped to the floor. Then the intense pounding began.

I received kicks to my face and stomach. The guards used

their sticks, and one of them broke some strange wooden thing and got the piece of wood and used it to hit every part of my body. All the time the officer was saying things like, "You sons of . . . you . . . We've been on the street since last night, you cursed . . ."

I told him, "God help you. Do you know why you have been out since the night? Or why you are fighting me?"

He got provoked and beat me harder. "You think you're smart, you son of a . . ."

"I'm not smart, I'm here for your sake, I'm an Egyptian, just like you."

Of course my words meant nothing. In the midst of his curses, my words were as meaningless to him as the TV show *The Recipe of the Day.*

After he seemingly got bored with beating me, he told his guards, "I want you to kill him like the other guy, or otherwise I will come back and kill you. And if you're hungry, eat him." He walked off. For the next ten minutes at least, I was pounded so heavily that I'm really surprised I am still alive. It's now 36 hours later and I swear that I feel the pain in every bit of my body.

The strange thing is that when it was all happening, I reached a point when I was not really feeling the blows I was receiving. I pronounced the testimony of God's oneness in preparation to meet my Lord. Images flashed through my head: What will my family think? What will happen to the movie I was not finished editing yet? What will be on the Facebook page created in my honor? I wonder if there will be a page called "We Are All Amr Salama"? More important, I wonder if the police will issue a statement that says I swallowed my iPhone?

I screamed at the guards, saying things like "I'm here for your sake. Do you have any idea why you are beating me? I have a cellular phone, money, a car, and I'm well off. We are here for you. So that you can eat and feed your children."

For some divine reason, they listened. One of them was really moved and pushed the others back and brought me a chair. He said, "Sir, will you be able to walk?" After a moment I said I would try. "Please go fast, before the officer returns. He will kill you if he comes back." I tried to run away fast, but the officer returned and thought I was escaping, of course. The guards acted as though they were catching me. So I received a second beating, which made the first one seem like a 1940s Disney cartoon.

Then the officer was distracted by the capture of another protester. Another officer approached me and asked what I did for a living. He saw my ID and told me to "run off quickly before the first officer turns his attention to you again."

I ran, and soon the pain started to hit me. Headache, dizziness, and blurry vision. My eyes began to tear nonstop. I was not crying, but maybe it was because my nerves were completely shot. And my voice sounded like someone speaking from the bottom of a well.

I reached the office of a friend of mine downtown. He took me in and brought me things to drink.

He left me alone for a while and I found myself sobbing intensely. I don't remember if I have ever cried like that before.

I was not crying from the pain, or even from the humiliation, or horror. I cried for one reason: because I found myself starting to hate Egypt. I felt that Egypt's security officers made me hate her. Egypt's government that treated us with injustice made me hate her. Her passive people—who were once passive—were not on our side and made me hate her. And the corruption, oppression, etc., etc. How can I do the great evil of bringing another life into this country? How can I convince a child to love this country, fight for her, and pledge allegiance to her?

I thought to myself, *Why don't I leave when I get the chance? Maybe my friend was right when he said, "Egypt's only future lies in emigration to Canada."*

Minutes later the voice of reason returned. Or not necessarily the voice of reason; it's a voice that has always gotten me into trouble. I reminded myself of my beliefs, the ones that I am writing this article to share.

I remembered that my sense of belonging to Egypt is not compulsory, but it is rather my own choice. I chose it because to belong is good for me, not good for Egypt. It is good for me to know where I'm from, and where this place is, this place that is mine, where my home is, and my bed, where I feel that I've arrived, and where I'm no longer waiting to go home.

I remembered that I must be positive about any place I choose to belong to. I must remain optimistic regardless of the circumstances. This place will become better because of me and the people around me.

Without belonging and hope, I would really rather die. If I continued to live without a worthwhile cause, I would be no better than an animal living to eat, sleep, and enjoy short-lived pleasures that will never nurture my soul. This was a final choice that I need not revisit.

Even if the world around me thought I was overly romantic or dreaming, I didn't give a damn. I would be happy with my choice even having suffered what I had suffered.

And I discovered that the most important thing in the world is that I know these things. To know why I was beaten, why I went out to protest, and to know that I decided to take to the street without the help of sophisticated slogans or complicated political demands and that I was beaten because I want a better Egypt. I want an Egypt without an eternal absolute authority for any of its rulers. I want an Egypt with a narrowing social divide. I want for impoverished Egyptians to receive a minimum standard of dignity and fulfillment of human needs. I want to know that my child, when I have him or her one day, will receive a good education and good health care. He or she will have hope and ambition no matter what, even if he or she wants to become president of the republic. I want a police force that protects the people, not one that will do to them what was done to me and to others, and what is being done at every police station and on every street. I don't want them to do what was done to the martyrs Khaled Said and Sayyid Bilal. I want for someone whose rights were violated to head to the police station trusting that he will receive his due and that the officer will want to serve the people and serve justice. I want officers to be found at police stations, not guarding presidential or ministerial convoys all day, and not suppressing a demonstration, and not working for the sole motive of protecting a regime that has failed to dignify the officer with a reasonable salary and decent life; the regime that has delegated the officer as the object of people's hatred.

I discovered when I was beaten that now I am less afraid and that I will participate in protests again and again and again. I now know that if I die, I will die a martyr and I'm sure I will be transported somewhere better.

I discovered that my oppressor did not know why he was beating me. He felt that his reasons were illogical even though he may have spent long nights trying to make sense of them. It is possible he sympathizes with me but is even more afraid of penalties and punishment than I am.

More important, I now know there is hope. There is not only hope to see Egypt following in the footsteps of Tunisia. More beautiful than I can imagine: there is hope for a place where I would wish to bring forth children who would lead a dignified life and who would share in its further development.

I do not claim to be a hero of any kind. I saw people who were more badly beaten than I was, and others who were arrested, and even others who died and whom I consider martyrs. But I spoke with the ones who are still living. Most of them are proud of their actions, their fear has diminished, the willingness to challenge has grown, and they feel they are right. Most of these people emerged from their detainment stronger, feeling that the sun will soon shine no matter how late it may be. They know that no matter how vicious the storm, the calm will have to follow soon.

Most important, I discovered that popular sayings often have very real meanings: "That which does not kill us makes us stronger."

800 Likes 192 Comments 138,025 Views

The group I was with followed Amr's group. We ran quickly after them and suddenly found Central Security attacking from the front. Police began subjecting everyone in the front lines to violent beatings, and I witnessed the first couple of punches that Amr received on the face. Hazem and I, along with dozens of others, dashed toward a gas station and climbed its staircase, only to find hundreds of guards lining up to block our passage. I tweeted our location and added that we were about to be attacked. The security forces started stomping on the ground and letting out loud cries, sending us a frightening message: they were about to charge.

But then, out of the blue, their commander ordered them to move away and clear the area. We did not understand why they stopped, but we didn't waste time asking questions. We moved swiftly into a side street and headed in the direction of Tahrir, hoping we had escaped. We learned later that the forces had received orders to halt all engagement with the protesters and to allow us to reach Tahrir Square. I will always wonder why. Someone, somewhere in the security hierarchy, must have given such an order.

The scene at Tahrir was one of the most enthralling I had ever seen. Enormous numbers of protesters—thousands, if not tens of thousands—covered most of the ground space in the square. This was when I realized that Jan25 had succeeded. It would be marked as a historic day for Egypt's opposition movement. Many incidents of beating and violence, as well as tear-gas bombing, had happened before we arrived at Tahrir, but things here were now calmer, further proof that top security officials had ordered the rank and file to stop engaging the protesters.

We could not believe our eyes. I began tweeting like a madman on my personal account, urging everyone to come out and join the protest. The numbers were increasing with every passing moment. Marches were arriving from many districts of Cairo, especially Shubra, Imbaba, and Boulak, drawing in more and more people. At Tahrir they were received with ecstatic cheers. The thrill I felt as these marchers arrived was like no other combination of happiness, relief, and hope that I have ever experienced.

When the young ultras arrived at the square, they began launching fireworks into the sky. One unified chant burst from thousands of lungs. One central, overarching, radical demand had captured the attention of every member of that critical mass in Tahrir: to rid our nation of Hosni Mubarak. We all yelled, *"EL SHA'AB YUREED ISQAAT AN-NIZAAM!"* (The people want to topple the regime!)

A significant portion of the protesters came from Egypt's middle class. When I asked people how they had found out about the event, many said, "From '*Kullena Khaled Said*'" or "From April 6's Facebook page." Others simply joined the marches that broke out in various places around the city. The march that came from Arab League Street was in fact the largest of all; it had started in several working-class neighborhoods and snowballed on its way to Tahrir.

The Internet connection was relatively poor in the square. No one knew why, but of course speculations pointed to security forces and their desperate attempts to control the crowd. It is equally possible, however, that the immense network pressure within an area of a few square kilometers was too much for the servers and they couldn't handle the huge amount of traffic.

It was important for calls for an even bigger protest on Friday, January 28, to begin right away. I tried but failed to access the Internet from my UAE cell phone. After many attempts, I finally managed to announce on the page: "General strike tomorrow and the day after tomorrow (Wednesday and Thursday) and a grand protest on Friday, after the ritual prayers." That was, to me and to many protesters in the square, the most fitting retaliation for Jan25.

Protesters continued to flow into Tahrir while I spent some time in a corner of the square talking with some others to Central Security officers. I wanted to try to have some genuine communication with them and at the same time work at resolving any conflicts that might arise between them and the protesters. One of the officers was exceptionally decent and approachable. He was hit by a stone thrown by a protester but appeared to accept that as a minor incident. I sensed his support for the revolution, which was obviously in conflict with his need to obey his superiors' orders. Our conversation was useful in understanding each other's different perspectives. He spoke of the risk of chaos and the outbreak of crime, while I spoke of freedom of expression and the right to oppose a corrupt regime. I never thought that this same guy and his soldiers would blindly follow orders later that night.

The way the day was unfolding was spectacular. At 9 P.M. I was extremely hungry, so Hazem and I walked to a famous kushari restaurant right outside Tahrir Square. Life off the square seemed to be as mundane as on any other day. It was as though absolutely nothing were happening at Tahrir.

We sat at a table with fellow protesters. They were Muslim Brotherhood youth, and we conversed while we ate. Back at the square, many young men and women were cleaning up the area in preparation for the sit-in that had been called by some protesters and political opposition members.

Communicating with the outside world had become a significant challenge. There were obvious network problems, and rumors were circulating that security forces had cut off communications from the Tahrir area. I was not prepared to spend the night at Tahrir, so I decided to return to my mother's home to change and then come back. A severe headache had kicked in – the result of lack of sleep and the eventful day. At home, I relaxed with a glass of tea and went online to invite all my friends to come out and join the sit-in.

It was then that I discovered that for the first time in Egypt, the government had blocked Facebook and Twitter. It was now clear that the security apparatus felt desperate. Facebook and Twitter could no longer be accessed through local Internet providers, so everyone started discussing and sharing alternative ways to access them. The regime's decision to block these two websites was a grave mistake. They did not realize that it was impeccable testimony to the strength of the protests.

On my way back to Tahrir I received messages that I should not return and that the square was "on fire." But I still went. There I found a horrific scene. The police had wanted to clear the square for the night, but the protesters had insisted on turning their rally into a permanent sit-in, asking everyone to spend the night. Finally the police started using force to try to clear everyone out. Security forces were chasing protesters, and clashes had erupted. The entire square was covered with a cloud of tear gas. My eyes burned terribly. The protesters were very angry, and many had lost their tempers. The very protesters who had earlier chanted "Peaceful, peaceful" were now stoning the security trucks and the Central Security guards in self-defense. The area had become a battleground. The security forces attacked with rubber bullets, water cannons, and more tear gas. Not even the journalists were safe from the violence; Al Jazeera's cameraman was injured by eleven rubber bullets. The security forces had one goal: to empty the square at all costs. They didn't want the sit-in to be seen by hundreds of thousands of Egyptians passing the Tahrir Square area the next morning.

It was a long night. The security forces finally succeeded in clearing Tahrir Square, winning a battle—but the war was far from over.

Just before going to sleep at 4:30 A.M., I wrote on the Jan25 event page: "Jan25 is not the end . . . It is the beginning of the end."

7

My Name Is 41

BY THE TIME PEOPLE went home, the Internet had been flooded with images and videos from Jan25 protests across Egypt. I had witnessed some of the most moving moments of my life on that day, and back at the hotel I had the opportunity to watch even more incredible moments that were captured and made eternal online.

One image showed a protester jumping in front of more than a dozen police officers, who were fully armed and wearing helmets and shields. This striking picture was taken while both the protester's legs were still in midair, as if he had broken free from the force of gravity. One video filmed with a mobile device showed a young protester carrying a sign and walking by a cordon of police soldiers. The young man was singing, in a very powerful and determined voice, the words of a famous Arab poet: "If, one day, the people desire to live free, then fate must answer their call. Their night will then begin to fade, their chains will break and fall." He kept confidently repeating the chant to the soldiers and then said, "Please memorize it!" Some of them laughed sarcastically, but many looked in awe at this courageous young protester.

The police had adopted numerous strategies in the hope of preventing protesters from reaching Tahrir Square. At one point they even started to use water cannons. One video showed an armored truck beginning to spray water while moving in the direction of a crowd of protesters. Seconds later the truck halted. A brave young man had decided to stand in its way. He put his hands on his waist and confidently stared up at the truck. The driver moved the vehicle a bit forward, but the young man didn't budge. The driver then increased the cannon's water pressure, but another protester quickly redirected the water cannon so the water went straight up.

I had always watched in wonder as the lone Tank Man stood courageously before an armored vehicle in the famous Tiananmen Square scene in China in 1989. I deeply admired how that Chinese man risked his life for the cause he believed in. It never crossed my mind, however, that I would one day see a fellow Egyptian performing a very similar deed. On Jan25, thousands of Egyptians had begun to act selflessly, and our Tank Man could have been anyone. (He remains anonymous to this day.)

These and other images and videos spread like wildfire online, particularly once the block on Facebook was removed. Rasd's page had reached a quarter of a million fans in around twenty-four hours, after many *"Kullena Khaled Said"* page members followed the link on our profile. As Jan25 came to an end, one thing was clear: the protest was not over.

The next morning, in the wake of such a momentous day, my primary emotion was anxiety. The turnout on January 25 had been higher than anyone had imagined, and the Egyptian masses widely supported the uprising. But where was it heading?

I continued to rally support for the revolution through my posts on Facebook. But this was no longer just about a Facebook page or the Internet as a whole. It had gotten much bigger than that. The people on the streets had begun to move at a faster pace than the political activists. The mob was now in charge, whether it was rational or not.

Access to Facebook was only intermittently available in Egypt on the night of the twenty-fifth and the morning of the twenty-sixth, so I prepared an online form to collect e-mail addresses of the page's members. If Facebook was suspended till Jan28, we would need an alternate means to communicate with as many enthusiasts as possible. A few hours after I posted the form, more than 20,000 members had filled it out.

My fear and anxiety rose even further when several "*Kullena Khaled Said*" members sent e-mail messages saying that State Security was arresting people whose names appeared on the list of attendees for the event on January 25. They asked me to delete the event invitation, which had already reached more than one million people. I thought hard before complying—these messages could have been posted by State Security officers who wanted to halt our momentum. But another post came from a member of the page who had been arrested because of a picture on "*Kullena Khaled Said*" that featured him (a Muslim) and his Coptic friend urging Egyptians to protest and take action against the regime. So I deleted the invitation. History is made on the streets, not on the Internet, I told myself. Protecting everyone who participated in Jan25 was more important than documenting it.

Yet after some more thought, I believed it was important to have a Facebook event for Jan28, one that would inform as many people as possible of the protest locations. After much analysis and hesitation, I decided to create a new event for Jan28. I logged on and created an event titled "The Friday of Anger: A Revolution against Corruption, Injustice, Torture, and Unemployment." I informed page members that the locations were to be determined on Thursday night.

I did not know how influential the event page would be, with the revolution now already on the streets, but I deemed it critical to do everything possible to spread the word. The event invitations were circulating at no less than three or four times the speed of the Jan25 invitations! I posted that Jan25 was certainly not the end of our protests and was the beginning of the end of Mubarak's brutal regime.

We don't want him . . . We don't want a president who kept his seat for 30 years . . . We don't want a president for life . . . We don't want a president who transformed into an infallible god whose commands are irrevocable verdicts . . . We don't want a president whose son will become president . . . We don't want a president who treats Egyptians as though they were enemies and orders them to be shot with rubber bullets . . . We don't want a president who named streets, squares, universities, and institutes after himself . . . We don't want a president whose picture I see at least twenty times each day when I interact with government agencies.

1,156 Likes 247 Comments 185,833 Views

I couldn't be more blunt on the page:

Down, Down with Hosni Mubarak . . . Down, Down with Hosni Mubarak

1,863 Likes 403 Comments 203,167 Views

State-owned media news coverage, along with that of many privately owned Egyptian TV channels, was incredibly biased against the protesters. A media campaign had clearly been devised by State Security officials to mobilize public opinion against Jan25. One very famous TV anchor claimed that protesters in Tahrir were attacking unarmed police soldiers while the innocent soldiers were voluntarily carrying fainting protesters to ambulances. The biased media outlets deliberately told the public that there had been injuries on the soldiers' side, but barely mentioned that protesters had been injured. *Al-Ahram*'s major headlines on January 26 were about protests in Lebanon, and the newspaper noted that some Egyptians had celebrated National Police Day by handing police officers chocolate and flowers in appreciation for their great efforts. Political analysts affiliated with the regime announced that the Jan25 protests had been driven by foreign efforts to cause chaos in Egypt.

Distraught, I called my friend Mona al-Shazly, the TV show host, and urged her not to be impartial when she covered the previous day's events on her evening show. We spoke about the police force's physical assault on the protesters. I had seen the police use rubber bullets and tear gas against tens of thousands of young Egyptians. In turn, Mona asked me to speak on her show, *Ten P.M.*, by telephone about the block on Facebook and Twitter, because the Egyptian regime was denying any involvement, claiming that the heavy traffic on both companies' servers had caused the collapse. I agreed immediately, since her show was a way to reach millions of viewers across the country.

Mona was very tense when she appeared on live television that night. She told me that she was under pressure from the Ministry of Interior not to mention the facts about the protests and to say instead that the number of protesters was only in the hundreds. But the protesters were pushing her to take their side and present the complete picture. She also told me that the ministry's insistence on compromising her journalistic standards might drive her to quit her job.

During my segment on the show, Mona called me on my UAE phone and asked me whether Facebook and Twitter had been blocked. Her guest in the studio, a well-known NDP figure named Hosam Badrawy, had said that the social networks had collapsed because of high traffic from Egypt.

"Yes, they were blocked," I asserted. "If they were not deliberately blocked, I would not have been able to access them using a proxy server, which I was able to do. These social networks are designed to handle hundreds of millions of users, not just four million." I went on to criticize the government's policies and its unwillingness to listen to the country's youth. I warned that ignoring the voice of reason would cause chaos. I began discussing the events of January 25, but Mona was quick to interrupt, thanking me and ending the call.

Right after I hung up, I found an e-mail from Mostafa al-Nagar asking that we chat online as soon as possible. The e-mail stated that there was an emergency. As soon as I logged into the chat session, Mostafa said, "Be very careful—the security forces are hunting down the page's admin. I was arrested last night in the protests. I was taken by State Security officers to an unknown location. They covered my eyes and I was severely tortured for hours, and they repeatedly asked me, 'Who are the real forces behind Jan25? What is the relationship between ElBaradei and the Muslim Brotherhood? Who is financing your movement? Who is the admin of the *"Kullena Khaled Said"* page?'"

I started to panic, and my heartbeat increased dramatically as he was typing. "What did you say?" I asked apprehensively.

"I didn't mention your name, don't worry," came his reply.

I felt relieved.

"I convinced them that I didn't know who the admin was and that he was probably living outside Egypt," Mostafa added. He explained that someone at State Security had continuously kicked him in the knees while he was blindfolded, saying, "Do you think you have defeated us? You think you've won? We allowed you to protest. This state is much stronger than you think and will not be affected by thousands or even tens of thousands of you protesting in the street."

After a few hours of interrogation, State Security lost hope that Mostafa would tell them something they didn't already know and released him. To this day, he doesn't know whether they believed him or not.

I was furious that my friend had been harmed, but I was also reassured because I had not yet been identified. Mostafa was arrested because he took part in the Jan25 demonstrations, not because he was my friend. Yet he and Mahmoud Samy, among others, knew my identity. Would I remain anonymous?

Immediately I decided to destroy the SIM card on my Egyptian cell phone, since my whereabouts could be easily traced while I was carrying it because the number was registered in my name. I felt that even remaining at the hotel was now too risky. I called a friend around midnight to see if he could put me up for a few days, but he did not pick up any of my calls. I tried another friend in a Cairo suburb, but the same thing happened. A third friend said he could offer me his small office in Zamalek, where there was only an uncomfortable couch to sleep on but there was a high-speed Internet connection. I immediately agreed and we decided to meet an hour later at the office, which gave him just enough time to get there from Heliopolis, also on the outskirts of Cairo.

Walking out of the hotel with my suitcase, I felt like a character in a spy movie. I evaded the taxi called by the concierge and crossed the street to get another. Two minutes later I was on my way to Zamalek.

To my surprise, my driver knew very little about the Tahrir Square events of the night before. The country had witnessed demonstrations of unprecedented magnitude, yet the regime was in denial pretending nothing had happened. The taxi driver's take on politics was the one shared by most of our eighty million Egyptians: There's no hope.

I had the taxi drop me off some blocks away from my destination. After glancing around and making sure the taxi had driven off, I started walking. My friend's office was a quiet, inviting third-floor apartment in a small building that overlooked the Nile. I was surprised to see a photograph of President Mubarak hanging over the meeting-room table, but then I laughed when my friend told me that it was intended to be ironic. At the time, Mubarak's picture was inescapable, prominently displayed in all government and public-sector offices.

It had been a long and tiring day. There was no way I could stay disconnected from the Facebook page, and the Internet in the office was down. My friend and I managed to connect through a backup network that belonged to another office in the same building. Relieved, I resumed my rallying for the following Friday and announced that the demonstration's gathering locations would be posted soon.

I couldn't get any sleep that night. I kept updating the page every ten to twenty minutes with videos and images from Jan25, and statuses and designs calling people to join Jan28. I eventually fell asleep on the couch around 11 A.M. on the twenty-seventh. I woke up three hours later, however, and the first thing I did was get on my laptop to check whether I had missed anything major. I closely followed progress on the event invitation page for the big day on Friday, which had already reached 500,000 Facebook users. I read the news and watched the talk shows from the night before on YouTube. The atmosphere was tense, and everyone was preparing for the demonstration. I received two lists of meeting points, one from Mostafa al-Nagar and the other from Ahmed Maher, after they had consulted with activists on the ground. I merged the two lists into one and published it on the page early in the evening. The list included names of many major mosques and churches across the nation. The objective was nothing short of a full-fledged revolution that would shake the ground under the regime. The driving factor now was the barbarity of the regime's response to the peaceful protests across Egypt over the past few days. We were not going to stop demanding respect for our dignity as Egyptians, especially now that we had begun to conquer our fear.

Facebook was blocked once again less than fifteen minutes after the demonstration venues were published, making it nearly impossible for protesters to locate the list. However, several activists had begun spreading the information via traditional means. I remembered the backup e-mail list that included 20,000 members, and I quickly set up a Google Group to send the message to everyone.

The revolution had transcended any single Facebook page, event invitation, image, video, or design. It was no longer a matter of tweets by activists. While looking at the photograph of Mubarak facing the desk I was using, I posted:

> The fact that President Mubarak has not come out and addressed the nation is the greatest proof that he and his government are terrified that what happened in Tunisia would be repeated in Egypt . . . Please, let there be no bloodshed . . . Our demands are perfectly just . . . We want a decent minimum wage for workers and monthly financial support for the unemployed . . . We don't want a president who will nominate himself for the sixth time, or one that wants his son

to inherit power . . . We want the emergency law to be annulled immediately . . . Habib el-Adly has to depart immediately . . . The parliament must be dissolved and new, fair elections conducted . . . We will continue until we either die as martyrs or our demands come to pass.

1,830 Likes 693 Comments 298,939 Views

Ordinary Egyptians everywhere were like boiling water that would soon bubble over the rim of the pot. Yet the regime completely ignored what was happening on the streets. Safwat El-Sherif, the NDP's secretary-general, addressed the people on television, saying that no member of government had fled the country because officials who had been sacrificing so much time for the sake of Egypt had nothing to be afraid of. He added that a true democracy never enforces the demands of minorities on the majority. His speech only intensified the anger of the youth. He not only failed to recognize our demands, but he was actually belittling the masses, referring to them as minorities who had no role in shaping the destiny of Egypt.

During the twenty-four long hours I spent in my friend's stuffy office, I ate nothing but fast food ordered from nearby restaurants and communicated with the world only by using the Internet and proxy addresses. My mother and siblings did not know where I was. All I told them was that I would spend the night at a friend's; I did not disclose his name. Meanwhile, my brother and I were exchanging messages through Facebook to avoid any possible monitoring of my phone calls. I yearned to get outdoors for a short while to breathe and get my body moving after sitting down for so long.

Earlier I had agreed to meet with two Google colleagues who were visiting Cairo from the United States. We had initially planned to get together on the afternoon of January 27, but given the rate at which things were moving, I suggested we postpone our meeting to later that evening and have dinner at Sequoia, a popular open-air restaurant surrounded on all sides by the Nile, which was a few minutes' walk from my friend's office. I remember enthusiastically talking to them about Jan25 and the Egyptian dream of change. About an hour and a half into our meeting, the restaurant's Internet connection suddenly stopped working. I had to leave, because I couldn't afford to be offline at this critical time. Sequoia seemed quite busy, almost as if no revolution was under way. My colleagues expressed their concern for my safety during the next day's protests and then we parted ways. Little did I know that this brief meeting would lead me to the most difficult experience of my life.

As I left the restaurant, I read news on Twitter that the text-messaging service was no longer working. At that point I tweeted, "Pray for Egypt. The government seems to be planning a massacre for tomorrow." I was walking down a dimly lit street. Suddenly three men jumped me from behind. They wrestled me to the ground. As one of them held me down, another clutched my legs with his arms, and the third sealed my mouth with his hand. With my voice strained from chanting at the demonstrations, I could just manage to yell for help.

"Shut up, you scoundrel!" came a harsh voice. "Don't you dare let out a sound." One of the men reported through a radio transmitter, "It's done, sir. We're ready."

There was no doubt they were State Security: their facial expressions, manner of speech, obvious experience with similar missions, and radio transmitter all told me what I was up against. They confiscated my computer, but I managed to keep my cell phone. They were trying hard not to make a scene. I was caught between two parked cars, and it was not easy for pedestrians to see me, especially since it was past midnight.

After a few minutes, a small car arrived. My heart was racing. Two of the men tightened their grip on me, and one of them said, "You are going to regret it if you scream." I surrendered and decided to stop resisting. I was now a hostage, and I knew what these men were capable of. It was best just to obey orders without speaking.

They violently shoved me inside the car. Two of the men sat in the back seat with me, one on either side, while the third sat in the front passenger's seat. They began giving me orders: "Empty your pockets, you ———." It was common practice for these people to use the filthiest curses with their victims. I gave them everything I was carrying. They handcuffed me and confiscated my glasses, watch, and cell phone, even the green wristband my wife had given me so that I would remember our children during the demonstrations and try to stay safe for their sake. The car drove off slowly, commencing its confusing twists and turns. It disoriented me, since the men had pulled my shirt up to cover my eyes and used my belt to wrap the shirt firmly around my head.

Many thoughts flashed through my mind during that ride. Above all, I was terrified that they had discovered that I was the admin of the Facebook page that had promoted the demonstrations on January 25. I thought about Nadine, the Egyptian activist in the United States whom I had asked to wait a week to announce my disappearance and relation to the Facebook page. I thought about my wife, who was so sad and so convinced that I was selfish, making decisions without thinking about my family's future. I thought about my kids and whether I would ever see them again.

I tried to calm myself down. But I could feel my heart beating faster every minute. It was horrifying to think that my laptop was on sleep mode and that all my applications were open, including the *"Kullena Khaled Said"* page, the page's e-mail account, and the e-mail group that I had developed to announce the venues for demonstrating. How could I possibly keep State Security from accessing my laptop?

The car finally came to a complete stop, and I was rushed into a building where it was clear that many more people were expecting me. One of the security officers quickly forced me to face the wall and ordered me to keep my eyes closed. He proceeded to take the shirt off my head and said, "If you so much as try to peek, you will regret it." I shut my eyes as tight as I could. Next he took off my undershirt, and then he blindfolded me once again with a thick piece of cloth, which he tied securely behind my head. He ordered me to remove my pants. Then he began threatening to rape me. "Welcome to State Security, you son of a b———," he said. "Here you will learn to behave yourself and mind your own business."

My legs began to tremble. The officer laughed as he yelled, "Get the stick for this son of a b———." I dreaded what I thought was to come, but he suddenly told me to stand up and put my clothes back on. He was laughing, and I was relieved. I understood later that this was in fact no more than a routine search during which the officers verified that I was not hiding anything.

The physical beating I received was not bad, but together with the verbal insults, it was enough for me to give in to their orders. I felt like a living corpse.

One of the men grabbed my arm and dragged me to a spot not so far away. "What is your name?"

"Wael Saeed Ghonim."

"Wrong!" he screamed. "From today onward you will forget this name. Here you are called 41. Okay, 41?"

"Okay."

"What's your name, kid?"

"41." I shivered.

Then came a series of orders. I was not to talk to anyone or even to think about removing my blindfold, and I was to do everything the officers told me to do.

I used to think that State Security was not stupid enough to arrest someone relatively well known, like me, because his friends and acquaintances might go on the airwaves and apply pressure for his release. Even if I was arrested, they would treat me relatively decently, I believed. But I was wrong. January 25 had changed everything. There was no concern for saving face anymore—not for State Security and not for the regime. This was a battle to the death.

We went down several steps, and I felt sure we were heading to an underground prison. The officers led me through hallways that reminded me of the car's route on our way there: a maze with many turns left and right. We entered a room that seemed spacious, based on the echoes I could hear. I was seated on a wooden chair and told not to move.

I asked for water. After the horror of my arrival, I was parched.

They brought me a bottle of water and I drank like a man stranded in the desert for days. I placed the bottle between my handcuffed hands so I would be able to drink some more later.

I heard footsteps approaching from outside the room. From their voices, I could tell there were three officers. Then the interrogation began.

First came my welcome note: "You're in our hands now, and there will be no mercy. Do you think that we are a bunch of kids, incapable of doing our jobs? We've been on to you for a very long time, and we know everything you've been up to. Did you think State Security was stupid? We only waited to find out who you were and who you worked for. Now we know everything."

Another officer, who sounded older, interrupted. "Listen, son. You're captured now and we could easily give you a taste of hell, but we are ready to help you come out of this safely, if you decide to cooperate with us by being completely honest. And of course, you know what could happen if you even tried to lie to us."

I knew how State Security interrogations worked. The officers would torture their victims until they got the information they wanted. I remembered my religious friend Yasser, who disappeared for several months during our third year of college. His parents told me he had been arrested. When he came out of prison, he had terrifying stories to tell about what he had experienced. He had been tortured in the most brutal ways in order to elicit a false confession. The security forces wanted him to admit that he played a leadership role in an extremist group which he in fact knew nothing about. Yasser told me at the time that State Security officers would test the strength of their victims by confronting them with completely false accusations. If you denied it strongly enough, they would believe you.

The third officer in my room continued to issue warnings. He said he knew I had been summoned to State Security before and that I was smart enough to know that a fast confession after my street arrest would guarantee gentle treatment.

"Look, we've kept an eye on you for a long time, but we still want to hear everything from you," he continued. "We want your whole life story. But be careful. If you lie, that means escalation, and escalation is not good for you."

I asked, "Where do you want me to begin the story?"

"We're obviously not interested in hearing about your childhood," he replied angrily. "Start from your trip to the United States. Why did you go there?"

I did not understand the real motive behind the question. I wondered if it was wise to mention IslamWay.com. Fear crept into my heart again, and I began telling the story as I had in 2007, during my first interrogation. I had made a point of remembering the story I had told back then, in case I was summoned again. Now I repeated that I had gone to the United States to get married and that I had met my wife through a friend. We got married and planned to live in the States, but after 9/11 we decided to return home.

The officer slammed his hand hard on what sounded like a wooden table. "Look, son, it seems that you want to play dumb. Don't underestimate us, I tell you. I gave you a chance to tell the truth, but it seems that you insist on lying. I will give you a second chance: why did you travel to the U.S.A?" he shouted.

I paused. Did he know more than I had told him? Was he bluffing? I couldn't possibly determine, and the risks were high. I decided I should come clean on this particular question. I told the story of developing the website and traveling to donate it to an American charity. There I met my wife. I included all the details I remembered in order to make my story credible. I had no idea how all this was received. A blindfold is a powerful tool.

Once I had finished answering, I heard a second, angrier slam on the table. The officer was even more impatient this time. "It's clear that you still don't want to tell the truth. Look, son, I'm warning you for the second time that if you don't tell us all the details that we already know, we will adopt other methods of extracting information."

I was quick to answer this time. "I really just told you the truth. I have no idea what details I could be hiding. I swear that I don't know what you're talking about."

"Okay, I'll make it easier for you, to show you that we know everything," he said. "Tell me what papers you signed while in the United States. And who did you sign them for?"

Again I did not understand the question. I had signed no papers that could get me in trouble. So I said very openly and honestly, "There were only two papers that I signed, my marriage contract and the contract by which I transferred ownership of IslamWay.com to the American Muslim charity for no return."

"You don't want to tell the truth," he said sharply.

I thought again of Yasser. He had faced false accusations, not just to test his resolve but also as a means to get him to reveal information unknown to his captors. They made guesses about events and announced them with confidence, waiting for him to break down and confess.

I swore that I was telling the truth. I had not signed any other papers, I insisted. I asked the officer to present any information he had so that I could either refute it or confess to it. After two minutes of angry debate he said, "Okay, I will let this question pass for now, but just remember that you're still refusing to tell us the truth – the truth that we already know. Now tell me about your online political activism."

I fought panic. Were the questions about America a bluff, only a test before getting to the heart of the matter? Or were they just fishing in the dark about both topics? I had no idea. I didn't dare tell the truth about my relationship with *"Kullena Khaled Said."* The risk of telling the truth this time seemed immense. Instead, I answered that my personal Facebook page had more than 1,500 friends and my Twitter account had more than 3,000 followers. I noted that I regularly wrote online against the regime and in support of Dr. Mohamed ElBaradei and the movements for change in Egypt.

"So that's all?" he said cynically. "It seems that you are craving our other methods."

I stuttered while answering swiftly, "I really don't understand. I swear to God that I love my country dearly and have no idea why you are doing this. I did nothing against the interest of my country."

The officer became impatient and his tone rose. "You have betrayed your country! People like you should be ashamed." *Was I already doomed?* It no longer sounded like a bluff on his part.

No one knew where I was, not a friend nor a family member, and no one had witnessed my abduction. I was a captive, and my interrogators could easily take my life without the least trace of incriminating evidence. I reminded myself that my fate was in God's hands, not theirs. I decided to be truthful about everything and pray for deliverance.

"I am the founder of the Mohamed ElBaradei Facebook page, and I am also the founder of the page *'Kullena Khaled Said.'* It was I who determined that January 25 would be the day to launch the demonstrations as a revolution against torture, corruption, injustice, and unemployment." I trembled as I spoke. I wished I could see their faces, to find out whether they already knew all this.

"So it was you who incited people against the Ministry of Interior for months. You think you are a hero, and to make political gains, you spread false information about a person who died from drugs. It is all clear now. You work for the enemy. You are a despicable traitor."

I broke into tears. "I swear that I love my country more than anyone can imagine. I only created the page because I want to see a better Egypt and better police, and I want to see the day when Egyptians reclaim their dignity."

My tears did not deter him in any way, yet he did not sound as angry as I feared he would be. "What better police? What evil do you think the Ministry of Interior is committing?" By reacting relatively calmly and rationally, he gave me momentary hope. No doubt his purpose was to explore my ideology and allegiances.

I spoke about abuse of the emergency law, State Security torture, and other crimes committed by police officers. I do not remember every detail of the conversation, but I remember clearly that my interlocutor was not the least convinced that someone could possibly want to fight injustice without a hidden agenda.

He asked about my relationship with Dr. ElBaradei. This was an important part of the interrogation. The questions were specific, and my captors were searching for clues to help them unravel what they were convinced was a master conspiracy against Egypt. I described my relationship with Dr. ElBaradei in all honesty. I told them that I had met him only once. I noted that the accusations that he was an agent of the United States and a facilitator of the war on Iraq, bent on destroying Egypt and spreading chaos, were outright defamation.

I was dumbstruck to find that these State Security officers seemed to have the same impression of ElBaradei as many ordinary Egyptians did. I defended ElBaradei and also explained that although I might not agree with all of his ideas, I certainly shared his dream of changing our nation. The discussion became largely intellectual at this point, and it lasted for more than thirty minutes.

"So if you are so patriotic and you believed that you were doing nothing wrong, why are you keeping your identity a secret? Why do you conceal your name as 'admin'?" asked one of them.

Gently I answered, "For two reasons: because I seek no fame or recognition, and because I feared this very moment. I knew I was attacking the apparatus that could arrest me and throw me in jail under any accusation whatsoever."

"It's clear you think that all police officers are a bunch of monsters."

"On the contrary. If you've been following '*Kullena Khaled Said,*' then you must know that I have repeatedly declared that I do not fight the many respectable officers, I only fight the corrupt ones. I even published apologies when I wrongly accused any officer."

He interrupted: "The problem is, you guys think you know everything while you really understand nothing. You are being used to destroy your country."

"Our country is being destroyed every day by the politicians who run it, steal from it, and spread injustice among the people," I quickly responded.

The high-minded discussion abruptly ended. The officers asked for the names of everyone who had promoted the invitation to demonstrate on January 25. I decided to give a partial answer and supply only names that they already knew. I explained that Jan25 began as a reaction to the events in Tunisia. It was a simple invitation on Facebook that unexpectedly spread like wildfire across the entire web and moved onto the streets. I swore, truthfully, that there were no master plans or strategies.

"Who determined the locations?" he asked.

"Ahmed Maher from the April 6 Youth Movement and Mostafa al-Nagar from the ElBaradei campaign," I answered.

"How did you communicate with them?"

I asserted that the two men did not know me and that we communicated via the Internet, despite the fact that this was true only for Ahmed Maher. I deliberately tried to keep Mostafa out of trouble, and I wanted my story to be consistent with his. He had said that he didn't know the identity of the page's admin. The officer was not the least convinced, but he was pleased that I mentioned names and he made a call to run a check on both men.

I knew the names of many other activists who had coordinated the demonstrations, but I hoped to protect those who might yet escape scrutiny: Mahmoud Samy, who coordinated the demonstration on the ground in Mohandeseen and whose role was instrumental; AbdelRahman Mansour, the second admin of *"Kullena Khaled Said"*; and others. I did not want to endanger anyone's life.

"Who is doing the writing on the page now? Is there another administrator?"

Once again I gave a partial answer, "Yes, it's an Egyptian friend of mine in the United States. Her name is Nadine. She has admin access and might have started writing on the page." He asked for more details about her and about how I came to know her. I told the story of having lost the page and explained that Nadine had volunteered to post her name as administrator. We met because she was a member of Dr. ElBaradei's e-mail group.

They did not believe most of my responses. They kept asking me, how could the demonstrations be spontaneous when there was someone like me—an employee of an American company, married to an American woman, and working with a collaborator based in the United States? It did not make sense. The only thing that did make sense to them was the conspiracy theory: I had to be an agent of a foreign plot. This could not just be a spontaneous uprising. Security officers did not believe that the Egyptian people were capable of change or even of demanding their rights.

"Who did you meet in Sequoia, the restaurant beside the Nile in Zamalek?"

Again my answer was swift and direct. "Matthew Stepka and Jared Cohen, two Google employees from the U.S."

My questioner interrupted nervously, saying, "Yes, so these are the guys you receive orders from? You have betrayed your country and become an agent of foreign intelligence. Your job with Google is just a cover."

He really seemed to believe I was a traitor. The shock was more than I could bear. How could he accuse me of treason? Me? Someone who loved his country so much that he would not accept citizenship in another? My voice reflected my categorical disapproval. "I will accept any accusation other than treason," I said. "I swear that even if you shock me with electricity to make me confess to betraying my country, I will not. I prefer to die with honor than to accept such a heinous accusation."

"This case is crystal-clear," he said. "Your situation is hopeless. You are number 41, and I do not want to tell you who numbers 42 and 43 are." He was probably hinting at Jared and Matt. "We will get all the information in detail. Lying will not serve you well."

It had never crossed my mind that State Security officers would consider me a spy or traitor. I swore that I had not betrayed and never would betray Egypt, the love of which was in my blood. Why would I? I asked him. "I receive a handsome paycheck from Google, where I hold a good job. I moved to Dubai one year ago, where I live in relative luxury, and my work in technology is known in many places across the Arab world. Why would I possibly betray my country?"

I could not believe what I heard next. One of the three officers, who sounded like he was the biggest believer that Jan25 was a conspiracy, confidently declared, "You were recruited by the CIA, and the Google job was the perfect cover, since Google is an American company. Your privileged life in Dubai was the reward for your betrayal."

I tried to be rational and convince them that their theory was simply illogical. "Google is a huge company—its revenues exceed one hundred billion Egyptian pounds a year. Over 90 percent of their revenues come from advertising to end users, and over 50 percent of that comes from countries outside the U.S. Any links to the CIA could easily destroy the company's reputation and result in its complete collapse. The majority of Google's shares are listed on the stock market, and it is the board of directors that makes the company's decisions, not the CIA."

My answer seemed convincing to one of the officers, who asked me, "How then do you explain the sudden appearance of these two guys in Egypt today?," to which I gave more rational responses.

I asked him, "If current events in Egypt are part of a master foreign plan orchestrated through the Internet, how would you explain that a CIA agent enters the country at this critical time and meets someone like myself at a busy public place to discuss the demonstrations? Is the CIA capable of toppling the regime with such conspicuous stupidity?"

Once again I swore that I had said nothing but the truth. Jared and Matt's visit had been scheduled since early November 2010. They wanted to meet religious figures and decision-makers to prepare for a conference in Ireland about the best ways to counter violent extremism. Jared wanted to invite reformed radicals from around the world to speak at the conference, to tell their stories and describe their change of heart. I was telling the truth, but the timing of the visit was as bad as it could possibly have been.

"What exactly is the nature of your relationship to Jared?"

I didn't find it strange that they specifically asked about Jared. His Jewish-sounding last name could raise eyebrows in Egyptian State Security, given the long-standing Arab-Israeli conflict. "I never met Jared before, and we only video-conferenced once, three or four months ago, when he joined the company as the director of Google Ideas," I answered. "We talked over the phone a couple of times and exchanged e-mails after that to plan for his visit, which was initially scheduled for last November but was delayed because of some other commitments he had. We work in two different departments of the company, and I haven't had to interact with him beyond this project."

"What did he want you to do?"

"He had asked me and other Googlers in Egypt if we could facilitate his meetings with key religious figures such as the head of Al-Azhar [the largest religious institution in Egypt and the Sunni Muslim world] and the grand mufti, among others, including Amr Khaled and Moez Masoud. I began helping him with the scheduling of these meetings, but when I became busy with the protests in the past few days, a colleague from the Cairo office took over."

"Does he know about your online political activism?"

"He knows that I am politically active in Egypt and that I participated in the protests of Jan25," I replied.

"What did you discuss in your meeting with him?"

"We talked about the situation in Egypt. I asked him what he thought about the Obama administration's position on our demonstrations, because I know that he worked for the American State Department in the past. He seemed to believe that the administration considered Hosni Mubarak a crucial ally and that the Americans would support demonstrations only so long as they remained peaceful and did not threaten to overthrow their 'friend,' as happened with Ben Ali of Tunisia. He seemed sure that the U.S. government would hedge to protect its long-standing interests in Egypt by supporting Mubarak's regime."

I felt that one of the three officers believed me and wondered if this was simply the old "good cop, bad cop" routine. But I had no means of knowing for sure. That officer asked for the password on my computer, and my heart skipped a beat. I did not want them to access my laptop as it had the Facebook page's e-mail account, which contained 8,000 e-mails from the page's fans as well as addresses and lists of more than 20,000 people who had participated in the demonstrations. There was also a long list of volunteers who did not know me but who provided regular support to the page. To the interrogators, the laptop would be a treasure that would lead them to many other demonstrators.

"I need my phone to get you the password," I said.

Understandably, this seemed a bit odd to them. "Why do you need your phone when we are asking for your computer password?" asked one of the officers.

"I use a security application that requires a unique one-time password. The application that generates these passwords is installed on my phone so I can get a one-time password every time I request it," I said. My plan was to try the generated password on my computer, knowing full well that it would not work, since I specifically used this application only with my work e-mail account.

I was taking a big risk. If they did not believe me, they could torture me. Yet I had heard one of them make a call to inquire about my phone. He learned that it was available and in one piece. They continued to ask for information. They wanted the password to the admin account for the Facebook page. I quickly considered my options and decided it wouldn't do much harm to give it to them, since the most they could do was close the page down. The revolution was already on the streets.

The interrogators tried to forge a link between Dr. ElBaradei and the calls to protest on January 25. I told them the truth, which was that ElBaradei had no involvement with the demonstrations, direct or indirect.

They started asking about the upcoming Friday. "Who named it the Friday of Anger?" one of them asked.

"I did."

"Who designed the logo for the event?"

"It was one of the Facebook page members. I had asked volunteers to design a logo, and this was the one I ended up choosing."

"Who determined the locations for the demonstrations?"

"I don't know exactly, but I got two lists, one from Mostafa al-Nagar and another from Ahmed Maher, and I merged them into one."

"What is ElBaradei's role in the Friday of Anger?"

"ElBaradei announced through his campaign's page that he would take part."

"From which mosque?"

"It's all announced online. But anyway, he will pray at Istiqamah Mosque in Giza Square, then join the march to Tahrir."

I was drained. It had been days since my last full night of sleep. I was longing for the interrogation to end so that I could sleep and forget this horror movie.

But something happened—the interrogators quickly ended the session and departed, right after one of them received a phone call. I think someone else whom they wanted to interrogate had been arrested. After all, they were racing against time before Friday's march. Everyone expected a huge turnout for the Day of Anger.

After the interrogators left, a soldier approached and said violently, "Stand up." He then led me to the right side of the room and ordered me to hold on to the wall. Then he told me to sit. "You will sleep here," he said.

Using my handcuffed hands, I tried to make sense of my physical space. My bed was a coarse blanket of wool on the floor right by the wall. A similar blanket was folded and tied to serve as my pillow, and a third was for keeping warm, or as warm as possible. I wished I knew when I was going to leave this miserable place.

I placed my head down to try to sleep, but my thoughts refused to slow down despite how worn-out my body was. How had they discovered—if they actually *had* discovered—that I was the page's administrator? How much more information did they have? Would they be able to access my computer? What fate awaited tomorrow's demonstrations? How much time would pass before my friends and family discovered I had been kidnapped? And my wife—how would she feel?

All these questions flashed through my head, with no answers. I was wrenched between optimism and pessimism, between hope and fear of the unknown. I prayed to God that my crisis would end without dire consequences.

After a short while a man came and ordered me to stand up. He asked if I suffered from any chronic illnesses or took regular medication. No, I said. He took my temperature and left. The soldier inside the room asked me if I had eaten dinner and I said yes.

The only thing I remember next is a guard yelling, "Get up, kid. Number 41, get up fast!" I complied, and he directed me toward the phone, because someone important wanted to speak to me. It was the more trusting of the three interrogators, the good cop. He asked once again where Dr. ElBaradei would be the next day. Once again I named the Istiqamah Mosque in Giza. He also asked for the other page administrator's name, and I again named Nadine Wahab. He hung up and I was led back to my blanket on the floor. I was dead tired and slept deeply for a few hours.

I woke up to find my guard asking if I wanted to pray Fajr, the dawn prayer. I said yes, and he led me to the bathroom to do the ritual washing, or ablution, before the prayer. I expected him to remove my handcuffs and blindfold, but he told me to relieve myself and wash without removing the blindfold or cuffs. When it became clear that it was impossible to perform the ablutions while my hands were tied, he removed the cuffs, leaving my blindfold in place. I prayed, asking God to comfort my heart, mind, and body.

It would be very difficult to convey the psychological torture that I suffered during my abduction. I could write scores of pages and yet not adequately convey the feeling. Over the next eleven days, with my blindfold firmly in place and no sounds reaching me from outside the prison, I had little sense of time. If it weren't for the daily prayers I was allowed to make, I would have had no idea of days, nights, or hours. But with this morning prayer, I knew: the Day of Anger had arrived.

Outside the prison, Egypt was boiling, as I later found out. On the eve of the Friday of Anger (now known as Jan28), Dr. Mohamed ElBaradei held an urgent meeting at his home with key figures from the Muslim Brotherhood and the National Association for Change and leaders of his presidential campaign. Everyone commended the Brotherhood's decision to finally offer official support to the popular revolution. An agreement was made for Dr. ElBaradei to join the next day's planned protest at Giza Square. Some people who attended the meeting got arrested right after leaving ElBaradei's house.

Then the Egyptian regime committed a fatal mistake. On the morning of Jan28, all communication in the country was cut off. All three cellular operators, Internet services, and short messaging services ceased to work. Little did the regime know that this was the single largest promotional effort possible for the revolution. Every citizen who had not heard of the uprising now realized that a major challenge to the regime must be under way. Huge numbers of people decided to take to the streets, some for no other reason than just to find out what was happening.

The ritual Friday prayer was ideal for mobilizing the masses and later proved to be the main reason behind activists' success on Jan28. Most Egyptian males head to the mosques in their neighborhoods every Friday. The only concern at the time was that State Security would attempt to influence the imams, as they had done in the past, to dissuade people from protesting. Some imams would follow orders out of sheer fear of the consequences of disobeying, while others might sincerely condemn the protests, particularly after the huge nationwide media campaign against Jan25.

A friend of mine attended the Friday prayer in Nasr City, a few kilometers from Tahrir Square. The imam dedicated the entire sermon to discussing Egypt's condition. He talked about corruption and how people had become extremely frustrated with the status quo. He said that it was time for the government to take real steps toward change. Yet he condemned the calls for protests and talked about hidden agendas and conspiracies to destroy Egypt.

Some young men started to become frustrated, and when the imam rhetorically asked, "Who is going to benefit from such suspicious calls to protest today?" one of the attendees angrily shouted, "Egypt!"

It is against Sharia to speak to an imam before he finishes his sermon, and so the imam ignored the comment and again asked, "For whose sake is this all happening?"

Now it was more than one person who replied, "Egypt!" Then an even larger group of youth started chanting, "Egypt! Egypt! Egypt!"

My friend quickly headed to the imam and suggested that he conclude his sermon and lead everyone in prayer, in order to avoid any conflict. The imam readily agreed. Right after the prayer was over, a few people started inviting the others to head to Tahrir Square. Around twenty people began to march together, and after no more than a few hundred meters they had grown exponentially in number, as similar small marches from the other mosques in the vicinity joined with one another.

Other imams, however, decided to support the protests publicly. At Mostafa Mahmoud mosque in the Arab League Street in Mohandeseen, the imam's sermon clearly focused on the importance of putting an end to injustice. Although he didn't directly invite people to protest, his sermon unequivocally carried a message of support. Some of the youth who were attending the prayer even clapped for him on more than one occasion. As soon as the prayer ended, a prominent activist for workers' rights stood up and loudly chanted, "The people want to topple the regime!" Hundreds more began to chant inside the mosque. The march that began in the Mohandeseen area numbered in the thousands, if not more.

Across the country, people marched from their mosques to the major squares in their cities, particularly in Cairo, Alexandria, Suez, Ismailia, Gharbia, and Sinai. The Egyptians who took part in Jan28 came from different social classes, genders, age brackets, and educational backgrounds. Elderly men walked side-by-side with teenagers and university students as they chanted together, "The people want to topple the regime!" An oppressive regime had suffocated them for thirty years, and they were ready to fight back at last.

The first weapons the Central Security forces and riot police used against the protesters were water cannons and tear gas, and the streets turned into combat zones. Despite chants of "Peaceful, peaceful," the forces continued their aggression. Younger protesters became angry and started to fight back. They threw tear-gas canisters back at the police. They threw stones. Then the police started randomly firing rubber bullets, which led to many serious injuries. But Egyptians' courage and determination on that day were legendary. They were transformed into an army of citizens fighting their attackers. People rushed to offer medical attention to the persistent youth who suffered injuries and food and water to the rest. In Cairo, huge mobs marched from different parts of the city despite police resistance and reached Tahrir Square. Dozens of people were killed in the process; many were run over by armored vehicles, and others died from birdshot. Hundreds were severely injured.

The lopsided fight continued from the end of the Friday prayers, at about 1 P.M., until the time of the afternoon prayer, around 3:30 P.M. At that point the Central Security forces appeared to run out of energy and ammunition. They were no longer capable of facing the raging population, so they stood passively. Some people, despite the aggression they had suffered, saluted the soldiers and officers as they marched forward. Some of the soldiers and officers cried. Many of them hugged the marchers and broke into heartfelt apologies. "Forgive us," they pleaded, "we had no choice. But we swear, our hearts are with you. We suffer just like you do."

Many angry Egyptians started to march in groups to the police stations in their neighborhoods, particularly in the poorer areas of the country. Other groups marched to the nearest NDP headquarters they could find. To many people, both the police stations and the NDP headquarters symbolized the injustice and corruption of Mubarak's regime. Even today controversy remains about exactly what happened when these groups arrived, but by evening several police stations in different areas of Cairo, as well as in Alexandria and other cities, had been set on fire. Some claim that the police started the violence, shooting rounds of live ammunition at peaceful protesters in front of the stations. In this version of the story, the protesters' reaction to these attacks was to take over the stations and then set them ablaze. Another version claims that when the protesters arrived, they immediately tried to take control of the stations. The police allegedly reacted by firing live ammunition at them, especially because they believed that the protesters were actually thugs who sought to free the prisoners held at the stations. Regardless of which story is more accurate, it is certain that many people were shot dead and that police forces ended up fleeing, abandoning their posts in many critical locations across the nation after the stations were set ablaze. NDP offices in many cities were also set on fire, including the main Cairo headquarters along the corniche, which still stands in ashes today.

At 5:30 P.M., an announcement came down from the military saying, "Due to the events in some of the nation's governorates, which included riots and violations of the law as well as incidents of looting, destruction, burning, and assaults on public and private property—including banks and hotels—the president of the republic, in his capacity as military ruler, issued a decree of enforcing a curfew from 6 P.M. until 7 A.M. effective today, Friday, until further notice." The armed forces, in cooperation with the police, would enforce the curfew and secure public and private property. Tanks immediately spread out through the streets, with four tanks allocated specifically to the state's Radio and Television Building. Military forces also took control of Suez from the police.

The army issued a statement calling on the people to stand up against vandalism and protect their interests and the interests of the nation. Everyone was asked to respect the curfew, which now extended from 4 P.M. to 8 A.M. Violent confrontations continued, particularly in Sinai, where the State Security building at Rafah (on the border with Gaza) was bombed.

Army forces continued to deploy tanks and troops everywhere. Two or three tanks rolled into every major square, yet protesters met them with joy. This attitude was primarily inspired by the Tunisian revolution, in which the army had sided with the people and not the regime. Many offered flowers to the troops and took pictures of themselves standing by the tanks and the soldiers. They also chanted, "We're all Egyptian. The army is ours," and "The people and the army are one hand," while persisting with the main demand: "The people want to topple the regime!" Yet protesters remained unsure about the army's actual position at the time and continued to send out the clear message that they were opposing Mubarak and his Ministry of Interior, not the Egyptian military.

Many Egyptians abroad were horrified by the scenes that were broadcast by the international media. They began to organize and took it upon themselves to respond to the various statements that Egyptian embassies had made to the foreign media, primarily about how Egypt was heading toward chaos and the continuity of the regime was the only way to ensure the stability of the country and the whole Middle East. The efforts of these Egyptian expatriates were pivotal in guaranteeing that the world saw the truth of what was happening in Egypt and, in turn, engendering immense international support for Egypt's revolution.

Right after the tragic events of Jan28, the admin of the English "We Are All Khaled Said" page, Mohammed Ibrahim, decided to utilize the page's growing number of members—10,000 new members a day were joining during the revolution—to promote protests in twelve different countries, including the United States, Canada, the United Kingdom, and France. The page also helped coordinate between the protesters in each country. This impressive show of solidarity not only bolstered the confidence of the protesters in Egypt but also heavily pressured the international community to end all support for Mubarak's regime. Despite being separated by thousands of miles, Egyptians all around the globe were uniting under the banner of freedom for their country.

Late in the evening on Jan28, an announcement was made that the president was going to address the nation. Right after midnight, Mubarak delivered his first speech since the protests had begun on Jan25. In direct response to the protests, he announced his decision to abolish his cabinet, which included Habib el-Adly, the minister of interior. The next day Mubarak appointed a vice president. This was an action he had resisted for years, and his choice fell on General Omar Soliman, the head of intelligence. He also appointed Marshal Ahmed Shafik, who had been the minister of aviation, as prime minister and instructed him to form a new government immediately and to initiate a series of reforms. Although these decisions weren't enough, many protesters saw them as a confirmation that they would actually achieve their goal: "The people want to topple the regime."

Inside my prison cell, I had no idea of any of this. I was stuck in my blind isolation, ignored by my captors for the entire day. At around the same time, AbdelRahman heard the news of Mubarak's speech from some officers at his army camp. He wrote in his notebook: "Congratulations, Egypt, we've defeated the dictator!"

8

The Dungeon

"GET UP, NUMBER 41!" The security officer asked if I wanted one loaf of bread or two to go with the beans (what we Egyptians call fuul) and cheese that were available for breakfast. I asked for one loaf, and he handed me the bread stuffed with fuul. I was quite hungry by then, and it was surprisingly tasty. After my less-than-pleasant arrival at Security headquarters and my interrogation, I did not expect the food to be any good.

Around twenty-four hours before Mubarak's speech, I was sitting quietly in my cell and could hear some guards speaking about their personal lives—salaries, holidays, family—without any mention whatsoever of the demonstrations scheduled to take place that Friday. Perhaps they had been ordered not to. But my curiosity reached its peak when the noon prayer was over. How was the day unfolding? Was there any glimmer of hope that I could get out of here? I had no clue.

Looking back, I find it quite ironic that neither of the admins of the page that had initiated the call for the Jan25 protests actually attended the Friday of Anger on January 28. AbdelRahman was no less anxious than I was to find out how events were unfolding. He had learned from a fellow soldier at the camp who had a radio and listened to BBC News that Tuesday's demonstrations had been very large and had sent a powerful message to the regime.

His military camp's leaders were growing concerned that something unexpected might occur within the camp as a reaction to outside events. On Friday morning, some soldiers had received reports that no visitors would be allowed into the camp that day, as a precautionary measure. Yet others denied that any such orders had been made. This led to confusion among the soldiers, including AbdelRahman, who was anxiously waiting for a friend to arrive with news. In the end it turned out that visitors were allowed into the camp, and AbdelRahman received his friend, who showed him a copy of the *Al-Shorouk* newspaper for Friday. The headline was about the tragic events in Suez, where many protesters had been shot dead on Wednesday and Thursday. AbdelRahman felt extremely frustrated at not being able to join the protesters.

In my cell I continued to measure the passage of time by keeping track of the five daily ritual prayers. Lunch was a plate of rice with vegetables. Dinner was a loaf of bread with halva. My time was spent trying to sleep, to think, or to overhear any activity outside.

That night I heard the sound of a television coming from a room that seemed to be on a higher floor, possibly right above mine. I could barely make out the voice, but it seemed that someone was giving a speech. (Later I would learn that it was Mubarak's television appearance on the Day of Anger. He tried to mollify the opposition by dissolving his cabinet, but mainly he denounced the protesters for threatening order.) After the speech was over I decided to take a risk.

"Who was speaking on TV just now?" I asked the guard, almost whispering.

"Shut your mouth!" he screamed. Then he dashed out of the room and returned with someone who I guessed was his superior.

"Get up, 41!" this man said in a powerful shout. I stood. As soon as I was up, he yelled, "Sit down, 41!" Perplexed, I did as I was told. "Get up." I stood. "Sit down." I sat. He repeated this cycle over and over. It became clear that he wanted to discipline me. Finally he said, "Did we not tell you to keep your nose out of other people's business?"

Quietly I answered, "Yes, you did."

"So why are you eavesdropping outside? It looks to me like you are intent on getting yourself into trouble."

"I couldn't tell what was being said, I swear. I'm sorry. I won't do it again." He left me and went on his way.

From then on I kept my interactions to an absolute minimum. I did not talk to anyone or request anything unless it was absolutely necessary: permission to go to the bathroom or to fill my water bottle. I did not want to endure any more humiliation. Physical torture scars your body, but humiliation leaves scars on your dignity, and they can take much longer to heal.

Among all security officers, it was the guards who dealt with me harshly and condescendingly. They exaggerated their insults to make everyone feel like a nobody, worthless. My prayers intensified. I asked God to release me from this morbid place.

On Saturday morning I was ordered to stand up and move to another room. I had no idea why I was moving. I simply complied. They took me to a room with other detainees. I heard the guards addressing them aggressively and abusively. I sat there for several hours. It was extremely cold, even after I wrapped myself in the woolen blanket. Later I was taken back to my original cell, which I came to appreciate. I never found out exactly why this happened. Perhaps they wanted to show me that there were other detainees, in the hope that this would intimidate me. Yet I found the fact that I was not alone reassuring.

Three days passed before my guard told me I was due for interrogation again soon. I felt a strange happiness: finally I was at least going to talk to someone, even if he was a State Security officer. The deafening silence and blinding darkness could effectively render a human being insane, and for an Internet addict like myself, who thrived on communication and whose phone and e-mail never ceased their activity, it really was unbearable. That same day I noticed something new in my lunch: a piece of meat. It was the first meat I had been offered since my capture. It seemed as if the officers were trying to upgrade their hospitality a little before my next interrogation.

The guard led me out of the cell by the handcuffs and pushed me rapidly toward the interrogation room. I heard the interrogator's footsteps approaching.

"Wael, how are you?" the interrogator greeted me. I recognized his voice. He was the most sympathetic of the three officers from my first interrogation. This was a good sign, I thought. It was also comforting to be called by my name for the first time in three days of detainment.

"I'm okay," I replied in a low, sad voice. He asked if I wanted anything to drink and I politely declined, but he nevertheless told the guard to get me some juice.

The interrogation was friendlier this time, without accusations. He was trying to convince me that he believed I was a patriotic Egyptian who had been deceived into assisting a foreign group working against the nation's interests. I swore that I had told them the truth the night they arrested me. Jan25 protests across Egypt were nothing but a manifestation of anger accumulated over thirty years. I insisted that my role in releasing this anger was negligible compared to the role of the regime in building it up in the first place.

The officer repeated many questions from the first interrogation, and I was not sure whether he was just looking for confirmation or whether there was something more. He even asked me for the password of the admin Facebook account that I had already provided. I felt intense relief: they had not yet attempted to access my accounts! I disclosed the Facebook page information once again.

"I need the password to your e-mail account," he added.

That was quite challenging. During the first interrogation the officers had focused on my laptop's password and I had felt relieved that they hadn't thought to ask me about the e-mail account's password. Now I had to make a decision in a few seconds, so I decided to give him the password, praying that Najeeb would sense that I must be in danger and try to locate the password in order to reset it.

I was still dying to ask about Friday. But I feared how the officer might react. There was little reason to risk cutting short the only decent treatment that I had experienced so far. I hoped he would tell me himself, but he did not.

After twenty or thirty minutes of questions, he asked, "Is anyone treating you badly?"

I hesitated for a long time before answering, afraid of the consequences. Then for some reason I decided to tell the truth. "Yes," I said. "I have been insulted and beaten a few times during these three days."

He responded sharply, "This is not supposed to happen, and I will make sure that no one mistreats you at all." I was very skeptical about this promise. I knew that humiliation was standard and that the guards could not have been treating me this way without the consent of their superiors. He then commanded the guard who had brought me to take me back to my cell, and gave specific orders that I was be treated particularly well. I was still a bit skeptical, yet relieved. I knew about the "good cop, bad cop" strategy, but I couldn't care less. I needed a good cop now more than ever.

In my room, when I wasn't lying down, I would sometimes sit with my back to the wall for several hours. This began to take its toll on my health. A year earlier, as a reaction to Dubai's high humidity, I had suffered a severe chest allergy that kept me home for more than forty days and gave me a harsh cough. Now, as a result of being in a weak state and leaning on one of my room's damp and moldy walls for so long, the allergy was making a comeback. My coughing was so severe that a guard actually made a call to ask if I could be transferred to a hospital. He told the person on the other line that he feared the coughing would kill me.

Around a day or so after my meeting with the "good cop," who I later learned went by the code name Rushdy, my guard ordered me to stand up because an officer wanted to speak to me. I did not recognize the officer's voice when he greeted me, but he sounded relatively young, probably in his thirties.

"Hello, traitor," he said.

"I can accept any accusation you choose, but not treason. I am not a traitor. I would readily die for the sake of my country," I said quickly.

"Kid, you have betrayed your country and caused a huge disaster. God knows whether this nation will ever get back on its feet."

I assumed from his words that Friday had been a decisive day and that the protests were ongoing. But I still did not dare ask.

He took my handcuffed hands and said he would place them on his thigh. He pushed them at one spot that seemed to have some kind of scar. "Do you know what this is?" he asked.

"No, I don't."

"It's a bullet wound. I got it in the nineties, when I was chasing a group of terrorists who attacked the homes of peaceful Egyptians like your family and mine. Do you know that one of my friends died a martyr during that battle? You and the rest of those kids protesting do not appreciate the value of this country, and you have no idea how much we sacrifice for the sake of this homeland. You don't put your lives on the line like we do."

The conversation's informal tone confused me. I was not sure how to react. Should I provide a rational response and tell him what I thought, or should I just play along and avoid any unfavorable consequences in case I upset him? I tried to be diplomatic. I explained that I was not opposed to a security apparatus that defended our nation against aggressors but that I adamantly opposed a police force whose primary mission was to protect a corrupt regime. I mentioned the rigging of elections, the use of the emergency law to crush political opposition, and other matters. His response was that although he didn't think the situation was ideal, he nevertheless believed that it was the lesser of two evils, since without this regime Egypt would quickly sink into an abyss of mayhem and political discord.

Neither of us found the other's logic convincing. But after he left the room, I felt victorious. I thought I had at least convinced him that I was no traitor. Perhaps now he would believe that I was only a good person who had done something that the authorities considered wrong. It would be an improvement.

The aftermath of the Day of Anger was deadly, though I wouldn't learn of it until I was released. By daybreak on Saturday, January 29, more than fifteen protesters had been killed, and some, including the families of those who had been martyred, attempted to break into the Ministry of Interior's building in downtown Cairo, near Tahrir Square. Live ammunition was fired at the crowds until the army formed a barrier separating the protesters from the building. Elsewhere, inmates at several of Egypt's major prisons tried to escape, and the police used live ammunition in various attempts to stop them. Prisoners at Wadi al-Natroun Penitentiary on the Cairo–Alexandria Desert Road rioted and attempted to escape, many with the help of relatives who had come to liberate them. Several thousand inmates, including a number of political prisoners, were able to break out. Security sources told Reuters that 8 prisoners had been shot dead and 123 others had been injured during their attempts to escape. While the regime's interpretation was that local opposition forces, in collaboration with some foreign powers, had plotted to free the prisoners, revolutionaries maintain to this day that this was in fact a tactic employed by the regime itself in order to create chaos in the nation and then lead public opinion to blame the protesters for it.

Egyptians flocked to grocery stores and bakeries to stock up on basic supplies. No one knew how much, or how long, the disruption of trade and transportation would affect daily life. Yet signs of hope emerged. In the absence of security measures and police protection, Egyptians everywhere joined together in an epic display of solidarity. "Popular committees," formed to protect districts and neighborhoods from looters and thugs, sprang up spontaneously across the nation. Residents of a neighborhood, a street, or an individual building would divide themselves into groups and take shifts to guard families and property throughout the night. They also formed checkpoints at key entrances and inspected the licenses and IDs of passing cars during the curfew hours. Religious Muslims and Christians protected each other's places of worship, in a mutual effort to prevent any outbreaks of sectarian conflict like the conflict Egypt had witnessed only a few weeks earlier. The protesting youth played their part as popular committee members by night; come morning, they proceeded to Tahrir Square to participate in the ongoing sit-in demanding the president's resignation, democracy, and social justice.

At Tahrir, the sit-in reached its peak in the days following Jan28. The number of tents increased many times over with the arrival of participants from every corner of the nation. The sit-in participants were organized and divided duties among themselves. The square was full of very creative and dedicated people. Some groups were responsible for cleaning litter from the square, while others took care of securing the entrances and searching people coming into Tahrir. Doctors who were protesting established a field hospital in one of the square's small shops to provide emergency first aid to those who were getting injured. Plumbers brilliantly converted a few of the now dysfunctional armored vehicles into public bathrooms for those taking part in the sit-in. Other people created a lost-and-found desk to help people locate their missing belongings. One guy rigged up a power-charging unit using a light pole in the square to help people recharge their cell phones and laptops. A barber offered his free services to anyone who was taking part in the protests. Performers, both professional and amateur, sang every night in Tahrir Square, and some hilarious Egyptian citizens were sharing jokes and funny posters. This incredible spirit helped the protesters to continue with the sit-in until their demands were met.

Many citizens outside the square also played a critical role, delivering food and medical supplies. They also handled traffic control, directing drivers through detours and around barricaded streets. Such an organic civic movement, to my knowledge, was unprecedented in Egyptian history. The revolution brought out the best in people. It successfully proved that a multifaceted society like Egypt's could easily unite when its members shared the same dream, and could do so with dignity. Everyone came together for the sake of the country.

In Tahrir, protesters were still chanting, "Revolution until victory. A revolution in Tunisia and a revolution in Egypt." A military officer even joined the protesters in the square, and the ecstatic crowds carried him on their shoulders and chanted, "Down, down with Hosni Mubarak!" Meanwhile, Egyptian television announced that Mubarak had ordered the new government formed by Ahmed Shafik to take all necessary measures to enhance democratic rights as well as initiate dialogue with opposition forces, especially young Jan25 protesters.

As part of its desperate attempts to conceal the truth from its own people, at noon on Sunday the thirtieth the regime withdrew Al Jazeera's license to operate. By the afternoon the satellite company Nilesat had also suspended the channel's broadcast, but Al Jazeera quickly announced alternative frequencies for its Egyptian viewers.

On Monday, January 31, the number of protesters at Tahrir Square increased further, and a call was issued for a million-man march the following day to pressure Mubarak into stepping down. The authorities reacted by shutting down the railroads to keep people from elsewhere in the country from reaching Tahrir Square and participating in the march. To them, it was critical that the protests not reach the million-person mark. Possibly under pressure from the regime, both Muslim and Christian religious institutions were largely unsupportive of the protests, and by then various leaders had even called on Egyptians not to take part. Yet the presence of independent religious leaders from both faiths in Tahrir Square from early on belied the government's attempt to co-opt the religious authorities.

Several activists issued a statement demanding a formal announcement from the army about its position on the ongoing events. Would it continue to take the side of the illegitimate president, or would it support the masses who had taken to the streets to demand the right to a decent life? The activists' statement also called on the Egyptian people to participate in marches toward the presidential palace, Parliament, and the state television headquarters on what was now called the Friday of Departure (referring to the departure of Mubarak), scheduled for February 4.

The armed forces' response was a clear statement of support for the protesters. Their leaders said they understood the legitimate demands of the people and would protect the people's right to peaceful demonstration. The army did not and would not use force against the citizens of Egypt. It was a signal moment in the revolution: there would be no bloodbath at the hands of the army.

Inside my cell, cut off from Tahrir Square and the political earthquake shaking Egypt, I tried to avoid thinking of my wife and kids. What was happening to me was already enough to cause the worst depression, and there was no reason to add fuel to the fire. But one night I could not help it. Special memories of Ilka and the kids started coming to me. I started remembering how funny Adam was, and how Isra was an artist who would never leave home without her drawing kit. I wondered what would happen to my family if I never made it back to them. I began to cry. I felt sorry for myself as I remembered my wife's words of farewell. She had said I was selfish. The tears began dripping from my blindfold.

"Why are you crying, 41?" asked my guard, sounding unusually genuine.

"I remembered my children."

"You have children?" He was surprised. "You look young. How old are your children?"

"My daughter, Isra, is eight, and Adam is three," I said.

"Where are they now?" he asked, in a caring voice.

"Both are with my wife in Dubai. I live there, but I flew in specifically for the protests."

"Don't worry, then. Your children will be safe in Dubai. God willing, it will be only a few days until you will be out of here." I was sure the guard was just trying to comfort me. No one could make any real predictions about the future.

Meanwhile, in Dubai, Ilka was becoming desperate to find out what had happened to me. She describes her own experience better than I ever could:

> Thursday, January 27th was the last time I spoke to Wael, so when Saturday night came, I felt that something was wrong. I told him to make sure to contact me daily, even if it was just an SMS stating "I'm okay." On Saturday night I called his brother Hazem in Cairo, and when he said that none of the family had spoken to Wael since Thursday, I told him to check the hotel he usually stays in while there—but they did not have a recent record of his name. I don't recall how many days or if it was just a day that went by before I received a call from his sister Mai, reassuring me that he was fine because his friend had called her and stated that Wael was with him. "Did you talk to Wael?" I asked. "No," she replied. "What is his friend's name and where does he live and why didn't you speak to him?" I snapped. "I don't know, but he is okay," was her response. I thought it was strange of her not to know the answers to these questions. When I pressed further, Hazem got on the phone and admitted that Mai was only trying to comfort me. It actually had the opposite effect. No one had heard from Wael for a few days during a very dangerous time in Egypt.
>
> During his disappearance, I trusted who he trusted and was glad I had paid attention to names he mentioned. I called numerous contacts in his mobile tablet, including Yonca Brunini, a senior Google manager, to seek help. Yonca and I would frequently contact one another whenever we had new information, or just to share our concerns and feelings. His friends in Egypt and the UAE were also alerting the media of his disappearance, and soon his face was all over the news. I felt that with Google and our friends and family behind us, using all outlets to search for him, the probability of finding Wael was on our side. Although I was using every outlet to find him, I also had to be realistic. There were reports of bodies being dumped in deserted areas and unidentified people abandoned in hospitals.

Aside from family, my support group was composed of my friends and Wael's friends. People I had never met before, as well as people I knew, were calling to ask me if I needed anything, and to give me any information that may have been helpful in finding Wael. I kept telling our children that their father had to stay in Egypt a bit more because of all the work he had to do. Isra knew her dad was supposed to be home a while ago, so I told her that he was late because he was helping many people there. The children were oblivious to his real predicament, because I did not watch the news in their presence and told people not to speak about Wael in front of them. I would go about my daily schedule with school runs, play dates, after-school activities, birthday parties, and other outings and then come home, lock myself in the bathroom, and momentarily fall apart. I teetered between feeling angry at Wael for being involved in what he knew could cost him his life and feeling hopeless and completely numb. I barely slept or ate and could not sit still for long. I had to be up, calling people, gathering information, checking on leads, and, most important, never remaining idle at any moment.

To me, leaving prison was like a dream now. The alienation, darkness, and silence were killing me. I fell fast asleep after my brief conversation, telling myself that things were getting better. The interrogating officer had asked the guards to treat me well, the officer who had come to talk to me ended the meeting with a more favorable impression of me, and I had spoken to one of the guards for the first time, forging a brief connection.

Sometime later the guard received a phone call from the interrogator. He woke me and asked me to get up quickly and speak to an officer.

The interrogator's voice came over the phone. "Your e-mail password is not working."

"Then one of my friends must have changed it," I replied, trying to keep the happiness out of my voice.

"And why would they change it?" he asked angrily.

"Because they are worried about me and they're trying to keep me out of danger."

"What danger?" he interrupted. "What are you hiding in the e-mail?"

"I'm not hiding anything, I swear," I said, "but it certainly has the names of people who planned to participate in the demonstration and who helped with the designs and volunteered on the page."

How had Nadine or Najeeb realized that I had disappeared so quickly? This question gnawed at me. It is true that I had been praying that one of them would miraculously find out about my abduction, but had my prayers really been answered?

I would later learn that Najeeb had noticed on Thursday night that I stopped posting on the Facebook page and on Twitter. He knew that even though these social networks were blocked, I could easily use proxy technology to gain access. So he got worried. He tried to call me on Friday and Saturday, without success. When nothing happened to ease his anxiety, he decided on Sunday that he must change the password on the "*Kullena Khaled Said*" Gmail account. I only gave Najeeb access to the Facebook page account; I had not given him or Nadine access to the e-mail account. He could not ask the company to make the change (Gmail belongs to Google), as he didn't want to disclose my anonymous admin identity to anyone. He tried using the "forget password" option, but even though he knew a lot about me, he could not answer the secret question. He noticed, however, that the password could be sent to another address, which I had entered as backup. This other address was my personal Gmail account.

Najeeb called my wife, who by now was extremely worried about me. He asked if I had left any personal computers at home, and she said yes. Shortly thereafter he went to my home in Dubai to try to access every computer and see if he could get into my personal e-mail account. He looked through three laptops that I had at home, but none of them had a password that he or my wife could guess. He then asked if I had left a cell phone. She got him my personal mobile tablet. That phone was password-protected too, but my wife suddenly remembered that Isra would know the password—she loved to play Angry Birds on my phone. Isra immediately entered the password on the mobile tablet. Najeeb got all the information he needed, in addition to a list of my friends' names and their phone numbers. He was finally able to change the password on the "*Kullena Khaled Said*" account, and then he began making as many international calls as he could to find out if anyone knew where I was. Angry Birds probably saved many Egyptians whose full names were in my in-box a lot of hassle!

With my e-mail safe, my greatest fears were allayed. When the officer asked me how he could access "*Kullena Khaled Said*" and manage its content, I once again gave him the username and password, articulating the letters one by one. At this point I was hoping that both Nadine and Najeeb had carefully followed the instructions and continued to update the page without changing its account's password. I actually wanted State Security to access the page. If the officers failed to do so, they would start to doubt everything I had told them, and I would suffer the consequences. Besides, their having access could do no real harm, since neither the page nor its admin account harbored any secret information.

As soon as the officer told me that he had managed to log in, I said, "Thank God—now you know I wasn't lying to you."

"I can't see the page yet. How can I see it?" the officer asked, ignoring my comment.

"In the address bar, add '/elshaheeed' to the Facebook.com URL," I responded.

"It worked!" he replied after a few keystrokes.

I told him that he was now the admin of the page, and I suggested that he click on the settings option if he wanted to make the page private and hide it from the public. I was hoping he would do so, since this would give Nadine confirmation that I was in danger. Furthermore, even if State Security ended up closing down the page, Nadine could retrieve it, since to Facebook she was the official page owner.

Nadine and I had agreed that my identity as owner of the Facebook page would remain a secret even if I was arrested. We had also agreed that Ahmed Saleh, an activist whom she knew well and who was a friend of both AbdelRahman Mansour and Mostafa al-Nagar, would update the page. Then, if I had been missing for more than a week, she would go public about my identity in the hope that it would mobilize pressure for my release.

The rapid pace of events drove home one of the key strategies I learned from the revolution: to achieve your vision, you need friends and communication channels more than you need plans. The world moves too fast for even the best-laid plans to hold up.

In an extraordinary irony, the officer never did delete the page. To this day, I still do not understand why State Security did not modify the page or delete it. Perhaps the officers worried that such an act would indicate that I had been arrested, possibly triggering an angry response from the protesters. Sooner or later, however, the public was destined to discover my identity as the page founder.

In the early morning of Tuesday, February 1, the new Egyptian minister of interior announced that the police force had returned to its original motto, "The police serve the people," in an attempt to reduce the protesters' fury.

That same day Egyptians answered the first call for a million-man march. Hundreds of thousands of citizens in Cairo and many more across the different governorates took to the streets with a single demand: Mubarak had to go.

Again the president appeared on television to address the nation. He asserted that he was not going to run in the next presidential elections, in September, and that he had no desire to remain in power. Yet this time he did one thing he hadn't done in his first speech. He made an emotional appeal: "I cherish the time I spent in Egypt's service. I defended the soil of this homeland during peace and during war. Egypt is my home. In it I was born and in it I shall die. History will judge me, as it will judge others. And Egypt will remain a trust handed over from the arms of one generation to the next."

The speech was hugely divisive. One camp accepted Mubarak's pledges of reform and thought he should stay in power until the end of his term. Others rejected his speech, deeming it a manipulative attempt to stave off the end of his regime. In part the divide was generational. Many parents began to pressure their children, asking them to return home. Some protesters did indeed relent and leave Tahrir Square.

As soon as Mubarak's speech was over, small groups of his supporters in different parts of Cairo went out into the streets. Their chants hailed the elderly president and attacked the protesters in Tahrir. Revolutionaries thought that these limited pro-Mubarak rallies had been intentionally orchestrated by the regime, to send the message that not all Egyptians opposed the president.

On the "*Kullena Khaled Said*" Facebook page, Ahmed Saleh, the new admin, continued to upload images from Tahrir that showed the spirit of solidarity between people of all ages, religions, and social classes. At one point, when he criticized the president's most recent speech, a storm of angry comments flooded the page. It was clear that Mubarak had won many hearts. The page's event "The Friday of Departure" had fewer confirmed attendees than another Facebook event called "I will not protest on Friday." That event had already attracted more than 140,000 supporters, while the one on "*Kullena Khaled Said*" only had 55,000. Still, Ahmed knew quite well that the revolution had already taken to the streets, and he was not misled by the Facebook figures.

The next morning thousands of people began rallying in support of President Mubarak at Mostafa Mahmoud Square in Mohandeseen. State media attempted to portray them as another million-man march. As the pro-Mubarak supporters approached Tahrir, clashes broke out between them and the revolutionaries. Observers began tweeting that the NDP had mobilized large numbers of its members to demonstrate at Mostafa Mahmoud Square and near Tahrir Square. Reports also circulated that businessmen loyal to the NDP had recruited some muscle to attack the protesters, paying four hundred pounds per thug. Protesters in Tahrir searched many of the thugs they captured and found NDP membership cards on some and police force IDs on others.

Around five hundred people gathered outside the state television building (near Tahrir), raising signs that said, "Yes to Mubarak for the sake of stability. Yes to the president of war and peace." Other signs said, "We will not be another Iraq," and "Whoever loves Egypt should not drown Egypt."

At around 11 A.M., pro-Mubarak rallies approached Tahrir Square. Participants in the sit-in found that Tahrir had been infiltrated by hundreds of Mubarak's supporters. The protesters stood together to force the supporters out of Tahrir and succeeded in pushing them toward the National Museum, at one end of the square. But after the afternoon prayer, around 4 P.M., the numbers of supporters increased, and riders on camels and horses came from the Pyramids area, led by thugs hired by the NDP and a few rich businessmen to clear protesters from Tahrir. The businessmen sought to speed up what they thought was the regime's victory, in hope of earning credit with Mubarak and Co.

This infamous "Camel Charge Battle" proved to many that the regime would stop at nothing. Clashes erupted on three major fronts inside Tahrir. The first and largest was near the National Museum and within the surrounding side streets. The second was on the road leading to Kasr el-Nile Bridge. The third and most violent was at Talaat Harb Street, where hundreds of thugs armed with knives attacked the protesters. The fate of the protesters looked bleak. Yet the tide turned when, unexpectedly, an army officer, Captain Maged Boulos, *fired at the thugs.* They backed off and fled into the side streets to throw stones. The protesters forced the thugs to retreat, captured many of them, and handed them over to the army.

By sunset the protesters had seized the horses and camels and delivered them to the army officers. The army designated a space at the entrance of Tahrir Street in Bab el-Louq where the animals would be kept. The protesters remained alert and prepared for further bloodshed. Indeed, more fights did take place as the night wore on.

Anyone who observed these battles knows that Egyptians are capable of the impossible if they stand united. Except for a limited number of now famous young activists, the people on the front lines were not political by nature. A lot of them were ordinary Egyptians from rural areas and slums. Many were Muslim Brotherhood members. Everyone chanted, "Defend Egypt. Defend your brethren and children. Defend your honor. Blessed be the martyrs, blessed be the martyrs."

By nighttime, Tahrir Square resembled a beehive of intense, organized activity. Everyone worked hard to protect the square against the waves of attacks. Some young people broke rocks into smaller stones and stacked them in piles to serve as ammunition. Others formed barricades with iron barriers and wooden planks collected from nearby construction sites. People put hard hats and even metal pots on their heads to protect them from stones. Older protesters patrolled the iron fences, striking them with metal bars to produce a loud noise that sounded like war drums.

The largest attack yet began at midnight from the side of the National Museum. Thugs stood on top of a few army tanks stationed nearby and pelted the protesters with stones and glass. The number of injuries grew, and a second makeshift hospital sprang up near the front lines of the confrontation. After about an hour, the thugs were forced back toward Abdul Moneim Riyadh Square and the protesters were able to fortify the barricades. The attackers began hurling Molotov cocktails at the protesters, but a group was able to climb onto rooftops overlooking the site and retaliate from above with stones and a few Molotov cocktails of their own. The attackers were taken by surprise and retreated further.

After the thugs retreated, they climbed onto the Sixth of October Bridge to throw yet more Molotov cocktails. Then the battle became deadly: snipers began targeting protesters and firing live ammunition from the top of the bridge. Martyrs began to fall, one after another, until a group of protesters was able to climb the bridge from behind the thugs and snipers and fight with them one by one.

Citizen journalism and social media played a great role in informing the world of the events in the square and many other areas across Egypt. Rasd, whose page was linked to *"Kullena Khaled Said"* on Jan25, had twelve admins who regularly collected information, photos, and videos from protesters. Their page quickly became one of the major news sources on the Egyptian revolution. More than 350,000 people joined the page to keep up with the minute-to-minute updates, something traditional media simply couldn't offer. This pro-revolution, crowd-sourced perspective was not limited to Rasd, either. Hundreds of other pages, Facebook accounts, and Twitter profiles were dedicated to disseminating the news. As a result, many Egyptians abroad were able to follow what was going on in Tahrir Square and accurately represent the protesters' point of view to their local media.

Moreover, the fact that regional and international media organizations such as Al Jazeera and CNN covered the events was key in ensuring the safety of protesters. Although the regime was collapsing, it remained keen on preserving a certain image to the outside world, particularly because it didn't want further pressure coming from the international community. It is truly a shame that the blood of an Egyptian was worth less to those tyrants than a picture or a video shot that exposed their barbarity.

Alone in my isolation, I couldn't begin to imagine that Tahrir Square was becoming a battlefield. While the pace of events was accelerating outside the prison, the minutes barely crawled by inside. My psychological state deteriorated, and I lost my sense of time. I forced myself to sleep, and whenever I slept I dreamed of my wife, my son and daughter, my family, and freedom. The escape method changed from dream to dream, but I always woke up trying to pull my hands apart, only to realize that I was still handcuffed. Once I felt myself waking up with my hands free, but I was only coming out of a dream within a dream, as in the movie *Inception*. When I finally awoke, I felt the old disappointment all over again. I continued to pray that someone would get me out of the hellhole I was in.

Just three days before the Camel Charge Battle, Ilka called Yonca

Brunini and told her that she was very worried about me. Yonca quickly contacted the VP of marketing, Lorraine Towhill, and asked her for advice about what to do. The situation was quite complex, because Google has a company policy of not getting involved in any political activities. Yet both Yonca and Lorraine were worried about my safety and wanted to try to help me.

After some brainstorming at Google on how to find me, a team made up of senior Googlers, Yonca, Lorraine, and Rachel Whetstone, the head of global corporate communications, came up with the idea of launching a Google advertising campaign, targeting Egyptians by using the company's search engine. These ads announced that I was missing and asked Egyptians to help provide any useful information that would help locate me. There was no mention of a suspected arrest, because the company didn't want to be confrontational with the regime, for my safety and the safety of other Googlers who worked in Egypt. The company's security team decided to take on the task of hiring private security companies in Cairo to search for me in hospitals and police stations. The goal was to find out exactly where I was and determine whether I was alive. Najeeb and AbdelKarim Mardini, an Egyptian Googler, chatted every day for hours, updating each other on the progress of the search. Many great friends at Google were worried and did their best to help me out.

Hazem, my brother, also started calling everyone he knew and asking them about me. Along with many of our friends, he went to police stations, hospitals, and even morgues to find any trace of me, dead or alive.

Meanwhile, back at the army camp, AbdelRahman Mansour had received a visit from his father on Friday, February 4. AbdelRahman was worried, as his dad seemed apprehensive.

"Do you know Wael Ghonim?" his father angrily asked him.

"Yes, he is a friend," responded AbdelRahman.

The father quickly interrupted: "Where did you meet him, and what exactly is the nature of your relationship?."

AbdelRahman replied apprehensively, "Why? Is anything wrong?"

"Wael disappeared sometime around Jan28, and many people believe that he got arrested. I learned from your brother that you were both involved in the *'Kullena Khaled Said'* page that called for the Jan25 protests."

AbdelRahman's fear mounted. On his way back to his small room, he started to panic about what could happen next. He was both worried about me and worried that I would spill the beans and get him arrested. To receive an accusation of treason while still in the army was no joke, and a person could be executed for it. This is exactly why I had made sure to delete any trace of communication between myself and AbdelRahman while I was still in Dubai. AbdelRahman, however, hadn't deleted our chat records, and he knew that this would constitute solid evidence against him. As a result, he would spend the rest of his days in the camp terrified that he might be arrested. Although AbdelRahman and I were in two completely different places physically, fate strangely saw to it that we were sharing very similar emotional experiences at the same time.

In my cell, I noticed that the number of guards had declined. At first the guards had changed shifts three times per day, and now they changed only twice. Once, one of the guards watched over me for a whole twenty-four hours, complaining vehemently to his colleagues about his need to go home. I also noticed that I could no longer hear the guards yelling abuses at the other detainees.

My coughing fits were getting worse. The medicine one guard gave me had no effect. I coughed constantly and aggressively.

Sometimes I began to believe that my continued detainment was a sign that the revolution was successful and achieving its goals. At other times, however, the fact that detainment had not yet ended made me fear that the revolution had been defeated. My mind wandered between these two extremes, with no evidence from the outside world to settle my doubts. This was driving me crazy.

Little by little, I was starting to lose hope that I would ever make it out. I began to think I had been forgotten. Hours and days passed when nothing happened and nothing changed. I heard nothing, saw nothing, spoke not a word. *I took part in a great cause and I must pay the price,* I reassured myself. *It does not matter if you die at thirty or forty or even seventy—what matters is what you did with your life.* From time to time I sang to myself the Arabic song "Dream with Me." AbdelRahman Mansour and I had often posted the lyrics on the Facebook page when we felt that defeatism was spreading.

At one point, probably during my seventh or eighth day of detainment, an officer unexpectedly entered the room.

"Are you happy with what you have caused?" he asked.

I identified his voice immediately. He was the same officer who had told me about his bullet wound.

"I have no idea what's happening outside, but I followed my conscience in everything that I have done," I replied.

"Despite all that's going on in the country because of you, I believe that you are a genuine guy. You made a mistake, without realizing its disastrous impact. Anyway, tell me, is there anything that you need?"

I could not believe he asked me that. It had been at least a week since anyone had asked what my preference was on anything.

"Is this a serious question?"

"Yes—if it's a reasonable request, I will consider it," he asserted.

I did not need time to think. The words burst out of me. "I'm dying to take a shower," I said. A rash had spread over most of my body; I had not showered since I had been detained, and it was not enough that I had been performing ablutions for prayer. Worse yet, my clothes were stinking and rotten from my sweat in the stifling air of an underground room. My pants smelled from the remnants of food that I had no way of wiping off properly each time I was forced to eat blindfolded and handcuffed. I had never imagined that a simple shower would be a far-off dream.

The officer seemed agreeable to the request and asked if I had any clean underwear. I explained that I had been arrested on the street and hadn't exactly had a chance to pack any spare clothing. He then ordered the guard to run up to the office and bring back a specific plastic bag.

"You are a good kid, Wael, and guys like you should not suffer like this," he said. "You did something you should not have done, but I will be good to you and let you have my clean underwear." Then he laughed. "Don't worry, it's clean and washed, and the undershirt is even new. I never wore it before."

I was grateful to the man, although I had no idea why he was doing this.

A little while later, when I entered the bathroom, the guards allowed me to close the door for the first time since my arrival. They even told me that I could remove my blindfold as soon as they left. The feeling was unbelievable: I could see again after days and nights of absolute darkness. But my eyes could barely withstand the shock of light. I started reconnecting with my body — I had lost weight.

I found a bucket of hot water and a big cup to pour the water over me, in addition to a soap bar that had also been donated by the officer. I began pouring the water on my rash-inflicted body and rubbing the soap hard on it. With every drop of water on my body, I felt reborn. In less than five minutes the water was used up. I dried my body using the underclothes I had been wearing before tying my blindfold again and announcing to the guard that I was done. He then led me back to my place.

In five precious minutes, my spirit had revived and my despair subsided. Even as I put my reeking pants and shirt back on, I felt energetic and optimistic for the first time in many days. When the officer came to check on me, I expressed my gratitude. I felt he well deserved to be thanked for his act of kindness.

Outside the prison, Tahrir Square was reviving as I had. Mubarak's speech had caused a lot of people to change their position, yet when the decision was made to try to kick the protesters out of the square violently, in the Camel Charge Battle, the revolutionary spirit surged again. People wanted to believe that the regime was changing, but the use of violence proved that nothing would change as long as Mubarak remained in power.

On Sunday, February 6, Vice President Soliman invited representatives of every major opposition group — including the Muslim Brotherhood, the opposition political parties, the National Association for Change, representatives of Jan25 youth, the leading NDP dissident Hosam Badrawy, and many others—to a round-table discussion. Some of the young activists (who later formed the Coalition of the Youth of the Revolution, which was made up of numerous youth movements and groups and was consequently very influential in the square) insisted that no dialogue was even possible before President Mubarak resigned. Nonetheless, representatives of many other groups attended, and so did my activist friend Mostafa al-Nagar. Mostafa later told me that the vice president insisted that President Mubarak would never resign, and the discussion produced no major breakthroughs. Soliman did, however, agree to facilitate the flow of medical supplies and food into Tahrir Square, and he made a show of listening to demands for free and fair elections.

At the end of the meeting, Mostafa pulled the vice president aside and asked him about me. Was I alive? Had I been detained? Would the regime agree to release me?

The vice president responded, "Is this young man one of the revolution's decent youth like you? Some say he has relations with foreigners and espionage."

"Wael is a decent Egyptian. He is a successful professional employed by Google. He is one of the many Egyptians sincerely calling for change. Everything being said about him is sheer lies," Mostafa replied. "And his mother has been in very bad health since his disappearance. We have no idea whether he's dead or alive."

The vice president promised to intervene for my release. According to Mostafa al-Nagar, he handed a paper to one of his aides and ordered, "Look up this young man and have him released immediately."

In my prison cell, it felt like time had stopped. Once again I started losing hope that anyone outside was going to save me. I wondered a lot about my children. What would Ilka say to Isra and Adam when they asked about their father? And what about my mother? She cherishes me dearly and always worries a lot about my brother, Hazem, my sister, Mai, and myself. I also prayed to God that my arrest would not complicate my father's sickness. Two years before, my father had been infected with a rare disease that deprived him of vision in one eye, and he was gradually losing vision in the other one. I feared that something would happen to him, especially since he lived alone in Saudi Arabia and had no one to console him.

In my despair, suicidal thoughts started to make their way into my consciousness. I did everything I could to block them out. Nonetheless, complete desperation was washing over me after days of lockup, without the least knowledge of what was happening outside.

Yet as my hopes were plummeting, a comprehensive search for me was being conducted all over Egypt. Friends, family members, and the world's largest search engine were all on my trail. Although not much information actually surfaced, my name caught the attention of local and international media outlets, and journalists started to follow my case closely.

Moreover, Dr. Hazem Abdel Azim, whom I had met as a volunteer in the ElBaradei campaign for change, was also quite worried about me. He was one of the few people who knew that I had founded the "*Kullena Khaled Said*" page. After hesitating for some time, Dr. Abdel Azim decided to go ahead and reveal my identity to a journalist, Muhamed ElGarhi, who published an article about my role with the page, generating even more media attention. This put immense pressure on a regime that respected its image much more than it respected its citizens' basic human rights. Ahmed Saleh, the other admin, was closely monitoring the news, yet he was unsure about whether he should mention my name on the page. After consulting with Nadine, Ahmed decided that it was in my best interests not to do that, especially because the story had captured the media's attention already. Yet it was ironic, as hundreds of comments on the page were either inquiring about me and asking for my release or accusing me of being a traitor and the instigator of the chaos in Egypt.

Activists whom I had never met or known were also making amazing efforts to find me. Israa Abdel Fattah, the well-known Egyptian activist, appeared on Al Jazeera TV and said that no negotiations would start until Wael Ghonim was free, and that he was one of the people who should speak on behalf of the youth. In Dubai, Najeeb called Mona al-Shazly, whose number he had found in my mobile tablet, and asked for her help. Mona would later comment that she came to memorize my full name as a result of searching for me around the clock.

Everyone's efforts must have paid off in the end. On the evening of Sunday, February 6, I was called in for interrogation. When I entered the room, the interrogator broke the news.

"Wael, our investigation shows you are innocent and not involved with any foreign parties. You will be released in a few hours."

I could not believe it. It is hard to describe how I felt. One can only learn the value of freedom when it is lost. Imagine losing your freedom to move, to see, and to use your hands, not for a few hours but for days, with no idea what might happen next.

"Thank God," I said, barely able to control my excitement.

"But, Wael," he interjected sharply, "there is a lot we need to discuss before you go. You are now someone eagerly anticipated by the media, and you have no idea what has transpired outside. You must listen carefully to what I will tell you now."

We spoke for hours. He began by relating the events of the days since I had been detained. He claimed that three hundred police officers had been killed and that Tahrir Square was occupied by imposters from the Palestinian Hamas movement and from the Lebanese Hezbollah. He claimed that the two foreign groups had helped use tractors to break into Egypt's prisons and release all the criminals. I also remember him saying that President Mubarak had been insulted by everyone: the protesters in Tahrir, Internet users, and even Barack Obama, who had delivered an offensive statement that "meddled in our domestic affairs" and had called on President Mubarak to step down from his position immediately.

I listened attentively to everything he said. All I wanted was to break free from my prison. It was a shock to hear that hundreds of citizens had died, including policemen and soldiers, during the protests. I felt awful about it, and broke into tears.

The officer continued to describe recent events. He said that President Mubarak had fulfilled all the protesters' demands but they still insisted on humiliating him in public squares and in foreign media. He asserted that State Security had information confirming that the protests were now being fueled by foreign powers and local extremists with the intention of spreading chaos in Egypt. This is when I began questioning his statements.

"Why doesn't Mubarak simply step down?" I asked.

"If President Mubarak relinquishes power, chaos will spread and sectarian clashes will break out. Hundreds will die as a result, if not thousands. But if he stays, the country will remain stable and the sought-after change will take effect according to the youth's demands," he explained.

His attempts to brainwash me went on for a long time. He defended State Security and denied that torture was a common occurrence. He insisted that people released from custody by State Security were too quick to make false accusations. I did not believe that, of course, because my own friends had been tortured by State Security officers. He claimed that the only true cases of torture were perpetrated by a minority of corrupt officers. He tried to convince me that most State Security officers were cultured and educated individuals. He, for one, held a master's degree in international law, he said, and there were many others like him.

Next he wanted to review what I was going to say when I emerged in public. This was his real reason for conversing with me for so long, I realized. His words carried a subtle mixture of friendliness and threat. He warned me of the potential chaos that I might cause if I inflamed public opinion by discussing any ill treatment. He simultaneously emphasized that he was not instructing me in any way and that I was free to say what I pleased so long as I accepted the consequences.

The truth is that his threat had an impact. I decided to conceal the details of my experience inside State Security for the time being. What I had gone through should not become the focus of public debate, especially considering the physical horrors that others had experienced when captured by State Security.

After more than four hours of a mostly one-way conversation, I made two requests. The first was to see the officer's face and learn his identity. The second was to call my wife and let her know I was fine. The first request was met with staunch refusal, with the justification that he could not violate orders. I countered that I just wanted to get to know him, and besides, if State Security was so full of trustworthy and patriotic Egyptians, why not reveal their faces? Why not come out into the light? He would not budge, so I dropped that point and lobbied for a phone call home. Allowing the phone call was not in his authority, he said, but he would see what he could do. He sent me back to my room.

My guard was no longer harsh with me. He congratulated me on the upcoming release, and it sounded like he was quite happy for me. An hour later, however, I had still not been able to call my wife, so I asked the guard to speak to the officer. I was consumed with worry about how much she must have suffered since my disappearance. I only wanted to let her know I was alive and soon to be released.

A few minutes later the guard returned, saying that the officer wanted to see me. My request had been approved. The officer switched into English to let me know that he understood the language well and would be paying attention to what I said to Ilka. They brought me my cell phone, which the officer used to dial my wife's number. He then held the phone close to my ear.

It was past two in the morning in Dubai, and Ilka was asleep. It took a while before she answered and asked who was calling.

"It's me, Wael," I said, in a hoarse voice heavy with emotion. She was suspicious, answering nervously that I could not be Wael. "I swear it's me," I declared.

"What's my mother's maiden name?" she asked.

I answered, and added, "She is the one that loves me more than you do," which was our recurring joke. Her doubt vanished in a flash.

"Oh my God, Wael!" she screamed. "Where are you?"

"I'm fine. I'm detained at State Security but will be out tomorrow."

She shouted in disbelief and joy, an explosion of pent-up emotion. Then she said, "You have no idea what Google did for you. You work for a great company."

"This is no time for Google," I interrupted hastily, though I was curious to hear more. "Tell me, how is Isra?" Our daughter was fine, she assured me, but she was fast asleep. "Wake her up, please," I asked. "I want to speak to her."

"No, I don't want her to know you were detained. She thinks you are away for work and very busy. If I wake her up, especially if I'm crying, she will worry." I readily agreed, though I was dying to hear my daughter's voice.

At this point the officer signaled that I should end the conversation. I told Ilka not to worry and that I had not been physically tortured. Per my agreement with the officer, I asked her not to mention anything to anyone. "I love you, and I truly miss you and the kids. I will see you all soon, God willing," I said. Then we hung up.

That one brief conversation lifted a mountain off my shoulders. All I wanted was for Ilka finally to go to sleep assured that I was safe, for the first time in ten days. The officer asked me if I was happy. I answered with a big smile: "Yes." I felt relieved. He said that my wife must be a great woman. He highly applauded her for trying to verify my identity before we spoke about anything, and for not waking up my daughter in order not to scare her. I thanked him as he asked the guard to send me back to my cell.

Back in the cell the guard asked for my clothes so he could wash and iron them before my release the following day. He gave me something to wear in the meantime. I tried to force myself to sleep before the big day, but it was impossible: I was so excited, and my mind was racing.

The interrogation officer soon called to see if I was awake. He could not sleep either, he said, and wanted to speak to me on a personal level. The guard led me to the interrogation room for another several-hour brainwashing session.

Once again the officer's purpose was to polish the image of State Security and pressure me to deny that I experienced any ill-treatment. He reiterated that Egypt was going through a very critical phase and that it would be wise to think very carefully before saying something that might put me and many others at risk. The meeting ended when I promised to seriously consider everything he said and, no matter what else happened, not to exaggerate my experience. He thanked me and praised me as a patriotic Egyptian. I thanked him and returned to my room.

Next came another milestone: the guard allowed me to remove my blindfold, after I promised to place it back quickly as soon as I heard anyone approaching. I wanted to see the space I had been locked in for so long. I had thought that the cell was large because of the echo I heard every time the guard spoke. To my surprise, when I finally could use my eyes, I saw that it was a small room, the size of a second bedroom in a small apartment.

The guards entered the room one by one, and I saw their faces and got to know them all. As a group, I had thought them to be nothing but indecent animals. As it turned out, they surprised me by appearing to be perfectly normal. Their faces were the same ones you see every day on the streets of Egypt. Some of them even carried the same hint of goodness you see in many Egyptian faces. It was hard to believe that these people could, under the right circumstances, turn into evil torturers. It was really mind-boggling. This only confirmed my conviction that these men were not my enemies; they were just tools of my real enemies. Egypt's real enemies were the leading figures of the regime, who had corrupted these people and stripped us of freedom and dignity for the sake of sustaining themselves in power.

I bid the guards farewell and even hugged some of them. I told them that I was not a traitor and that I loved my country. One of them said that even though he was not supposed to say it, he appreciated knowing that Egypt had people like me. He told me that he had been inspired by all that was happening, and he ended by saying, "Pray that I find another job and leave this place." These strong words made me think about how he was imprisoned by State Security just as I was. I promised him that I would do that, and I asked him to start believing in change, so that his children would live a much better life, one that their father should have had in the first place.

I will never forget the officer who gave me his underwear and granted me a much-needed shower. On that final day, he entered my room and engaged me in conversation, blindfold off, which was a bold violation of orders. Our conversation resembled my previous two brainwashing sessions. It was part of an orchestrated attempt to influence my future public statements. I thanked him again, this time with laughter, for the shower—needless to say, having a shower is a basic right for detainees—and I told him that I could not believe I was wearing his underwear. His underpants featured black polka dots—hardly what you would expect from a tough State Security officer with a bullet wound in his thigh. He laughed and pulled up a piece of the underpants he was wearing to show that they had the same pattern. We laughed some more.

The hours remaining after that encounter went by very slowly. The guards brought back all of my belongings. My wristwatch now had a broken band, and my glasses had been smashed. My wallet was intact—and was missing nothing, not even the coins. They handed over my green wristband, which I had worn ever since the beginning of the revolution, my laptop, and my phone. I signed all the necessary papers, including a form that provided all of my personal information and contact details—the very information that I had once worked so hard to keep secret. They took a few mug shots, and then, at last, they placed the blindfold over my eyes one final time and led me out of the building. I felt the outside air and smelled Cairo for the first time in eleven days.

9

A Pharaoh Falls

At about 7 p.m. I was led to what seemed to be a minibus while I was still blindfolded. I sat in the center of the back seat, and I could clearly recognize the voices of the two men seated on my right and left respectively. They were two of the interrogators who had questioned me throughout my detainment. As we drove, it was clear that the driver deliberately made many turns to disorient me, just as he had after my arrest eleven days earlier.

My family was waiting to see me, since various media had confirmed that I would be released on that day. As I was getting ready to leave State Security, an officer who had taken part in my interrogation told me that he had received orders to escort me to the office of the new minister of interior, who was keen to meet with me for a few minutes before I returned home. As soon as we reached State Security's official headquarters in Nasr City, the officer said it was time to remove my blindfold once and for all.

I took the blindfold off and looked around. There wasn't much light, which was great for my eyes. As with the guards, the "good cop," Rushdy, and the other officer, whose name I cannot remember, did not appear stereotypically evil, the way State Security officers are said to look. In their faces I could not sense a taste for sadism and torture. They did not seem miserable or cruel. Instead, the faces I saw could have belonged to any young middle-class Egyptians.

"Congratulations, Wael," said Rushdy, looking at me with a soft smile.

We went through the gates into the building. Security measures seemed relatively high. Many soldiers, weapons firmly in hand, surrounded the building. Rushdy and I took the elevator to the top floor, escorted by an officer who appeared to be one of the minister's personal guards. When we got out of the elevator, Rushdy bid me farewell, saying that this could be the last time we would ever meet. I reflexively squinted, not having seen daylight since my capture.

The minister's office was about three times as large as an average Egyptian family's living room. There was a desk with several chairs facing it, a meeting table in a separate section, and another section set up as a reception area that could accommodate at least six people. The minister was engaged in a phone call, which he ended quickly after I entered the room.

"Hello, there. How are you, Wael?" he greeted me. I returned his greeting as politely as I could, but I was filled with anger. Suddenly the feeling of helplessness that had ruled over me for eleven days somehow became empowering. I noticed two people sitting in the room. I recognized one of them, whom the minister introduced as Ammo Hosam Badrawy. ("Ammo" is a local title used by youngsters to address adults; it is similar to "uncle.") I had never met Badrawy before. All I knew was that he was a senior NDP official and a member of Parliament and that he was considered a reformist within the party.

"You look upset," said the minister. "What is it? What's bothering you?"

In a risky show of defiance, I said, "It's the picture hanging above you that is bothering me," pointing to the portrait of Hosni Mubarak mounted in an elegant frame on the wall behind the minister's desk. It was routine practice for all government offices, especially security offices, to display the president's image, to glorify him and demonstrate their occupants' allegiance, almost as if he were pharaoh.

The minister, Mahmoud Wagdy, who had been appointed to replace Habib el-Adly, smiled and tried to assuage my anger. "The president has really upset you, obviously."

My response was quick and direct. "I don't know him personally for him to upset me. The president is employed by the people and you treat him in a way that has transformed him into a god, worshipped and obeyed by everyone. Why do you place his picture everywhere?"

"Please be aware, Wael, this is a man who's over eighty years old — old enough to be your grandfather. It is impolite for you to talk about him like that. The fact remains, he is president of the republic and the leader of our nation."

To me, the point of the revolution had been to eradicate such tired thinking, used to justify the wrongful practices of the regime. I told him that I respected my elders as long as the elders earned my respect, but that someone who does not respect his people does not deserve the respect of the people. I was starting to lose my cool, but he interrupted calmly again.

"It appears you are a true rebel. I'm also a rebel, and I want you to ask about me when you leave. It was Habib el-Adly who expelled me from the Ministry of Interior, and I am back now because of all of you. God willing, the situation will improve significantly." He then pointed to Dr. Hosam Badrawy. "Dr. Badrawy, say something."

"Mahmoud *bey*," Dr. Badrawy replied, addressing the minister with the Arabic word for "sir," "the youth have a language that differs from ours. The notion of respecting one's elders is not exactly the same for them as it was for our generation. Their generation has different considerations. I experience this difference with my children."

Then he turned to me and said that his daughter insisted he deliver her greetings to me because she was at Tahrir Square, thanks to the Facebook invitations. "Look, Wael, no one can deny that what the young rebels did was great and that your intentions were good.

Things are changing significantly now. A few days ago, the president appointed me to become the secretary-general of the NDP, and I accepted on the condition that all previous, notorious party leaders must be replaced during the coming period."

My anger took over and I interrupted. "The NDP changing? What are you talking about? I hate everything related to this NDP, which has ruined life in Egypt. This party will only continue to exist over our dead bodies. I do not want to see the NDP logo on the street or anywhere in Egypt again."

Hosam Badrawy tried to defuse the situation with an extremely polite manner and decent language. After a brief give-and-take, the minister said, "Wael, I am really glad to have met you, and as I told you, the country will change and so will the Ministry of Interior. Dr. Badrawy was one of the people behind your release from State Security, and he came especially to drive you home. This should prove to you that the NDP is really changing."

His smile provoked me. "I have nothing against being driven home by Dr. Badrawy in his personal capacity. He is an Egyptian figure about whom I have heard only good things. But I refuse to be driven home by the secretary-general of the NDP, even if he was the reason for my release."

Dr. Badrawy smiled and said, "Let's go, Wael. I came because you are like a son to me and I was happy to help with your release. The NDP has nothing to do with the matter. You and the rest of the youth are Egypt's great asset. We will always move forward as long as there are young people like you."

I bid the minister and his assistant—who had not spoken and had not been introduced—farewell and left with Dr. Hosam Badrawy. On our way downstairs, I started to sense the power of what the youth had done at Tahrir Square. The minister of interior used to be the pharaoh's hit man, and now here he was, a regular person open to criticism.

We started our journey from Nasr City toward the other side of the Nile, where my family home was, in Mohandeseen. Soon after we got into the car, Dr. Badrawy received a call from my brother, Hazem. He handed me the phone, and when Hazem heard my voice, he shrieked with relief, telling everyone around him that he was talking to me. I returned his greetings and said I was on my way home.

On the way, Dr. Badrawy was trying to convince me that Egypt was undergoing real change for the first time. He had always been displeased with the ruling party and was considered a loyal dissident. He hoped to instigate real reform from inside the NDP, he said, since external attempts seemed largely futile.

I interrupted. "NDP members should be ashamed of themselves. What change are you speaking of? What party?" My tone of voice rose as I spoke. "The same party you say will reform itself is the reason Egypt was corrupted and the dreams of its youth were killed. It is the reason Egypt is miserably lagging behind and has no status in the world. It was the reason I disappeared for eleven days without seeing the light. My parents knew nothing about me. This party will all go into the garbage dump of history." I was quite emotional at that point. "The reformists like yourself also shared in the damage through your presence in that party."

Dr. Badrawy appeared severely affected by my words. His head drooped and I sensed that his eyes were tearing, though he tried to hide it from me. I was crying and continued to rant. Eleven long days' worth of anger burst out of me.

The car pulled up in front of my family's house, where, to my surprise, a crowd of journalists and cameramen were waiting. I hid my face from the cameras as I tried to soften my tone and finish what I was saying. I reiterated to Dr. Badrawy that I had nothing against him personally and that I appreciated his efforts to have me released. I decided to give him a final piece of advice before exiting from the car: "If you ask me, Dr. Badrawy, I would say, resign from the party. Do not stain your reputation any further. Resign. This is the right thing to do. And it will add pressure on the regime to start enforcing real change." I then thanked him and said goodbye.

Many reporters had been waiting outside the house since the 4

P.M. announcement of my upcoming release. It was now 8 P.M., but they were all still there to get a statement. I decided to try to hide my face with the laptop and make a dash for the door, but some of the reporters and cameramen were able to run after me and take photos of my homecoming.

As soon as I reached the third floor, I heard screams and whistles. My mother was with the rest of our family members — my maternal uncles and aunts and their children — in the apartment, all waiting for me. I rushed to her. Her tears overflowed as she hugged me, as though she had reconnected with me after years of separation. My sister, Mai, who had suffered a serious breakdown during my detainment, also burst into tears. My brother told me that our mother, known for her strength, had not cried once until he gave her the good news that I was alive and had been set free, whereas my father had broken down a few times when calling to ask about me. Yet my mother was in shock the whole time, and so I told her I was back now and she didn't need to be afraid anymore. Then I spoke to my father on the phone. I assured him that I was unharmed and that we would keep fighting for Egyptians' rights.

I wanted to speak out to the local media and share my thoughts with as many Egyptians as I could. I was careful to select a very popular show watched by a diverse Egyptian audience. Without thinking much about it, I called Mona al-Shazly. She was very relieved that I was out and unharmed.

"You have no idea how worried we were about you," she said. The earnest quality in her voice made me feel that she would probably be objective if she interviewed me.

"If I am to appear with you tonight, Mona, then I have two conditions," I said.

"Of course — you've earned the right to set conditions," she said. I could sense her smile through the phone line. "What are they?"

"The first is that I want to talk freely. I don't want you to lead the conversation. I want you to let me say everything I wish to say. And the second condition is that, since this interview will be really big, I want a million pounds." She seemed shocked at my request. "But I don't want it for me," I continued. "I want you to promise to pay one million pounds to the families of martyrs."

Mona accepted the first condition and then asked if I wanted to supervise the distribution of the funds myself. I said, "No, I want you to promise me your network will handle the process or donate the money to a charitable organization." She asked for time to negotiate with the managers of her network. I promised her that I would appear exclusively on her show. Minutes later, she called and asked me to head for the studio. The network had accepted the deal.

I had no idea how many people had actually died during the events and I had not seen any pictures yet, but I suspected that many of the victims were married and came from the working class. Surely their families would be going through difficult financial times, and our political situation would make it impossible for them to receive adequate psychological and financial attention.

I accessed the Internet for the first time since my arrest and, to my delight, found that *"Kullena Khaled Said"* was being updated regularly. I assumed that Nadine had granted Ahmed Saleh admin access to the page and he was posting on it, according to our plan. I also discovered that the number of page members had increased by over 100,000 since my arrest.

There was no time to write any notes or posts, but I decided to write a personal message for the first time ever. It revealed my identity without mentioning my name.

> Thank God . . . I am back . . . I have not changed, believe me . . . I love my country . . . By God Almighty we will change it.
>
> 27,991 Likes 27,230 Comments 1,883,138 Views

I did not want anyone to spread rumors that I had been brainwashed by State Security or that my perspective had changed after detainment. Such rumors had occasionally been started by State Security, and other times political activists under heavy pressure had had real changes of heart.

I caught a glimpse of some of the members' comments, most of which reassured me and gave me hope. I had been used to occasional attacks from members before the revolution. Now what touched me the most was their solidarity.

> Wael Ghonim, for the first time in history, the regime arrested you and the people freed you. Congratulations. Egypt loves you and thanks you.

The number of followers on my personal Twitter account had skyrocketed to over 30,000 from just over 4,000 before the revolution. My Twitter mentions flooded with support messages, "welcome back's" and congratulations. I tweeted:

> Freedom is a blessing worth fighting for.

Mostafa al-Nagar, Amr Salama, and Mohamed Diab arrived at my house. I had invited them over because I was eager to hear updates from my activist friends about what had happened during my imprisonment. I needed their input to formulate my message for the TV audience. Mostafa expressed concern that I might appear on TV without being adequately aware of what had been happening. His advice was to wait until I had been to Tahrir Square and met the people there. He also felt that I had spent too much time talking to State Security officers—activists had been known to change their opinions after being subjected to intensive brainwashing.

I was indeed under the influence of the hours-long brainwashing sessions that I had endured. I was feeling sympathetic toward my captors. Nevertheless, I insisted on going on the air on the same day as my release because I was charged with honest emotions about Egypt that I wanted to express. I was aware that the decision represented a risk, but I relied on my personal conviction that I was sincere in my endeavors and that accordingly, God would not let me down.

After our brief conversation, we all got in the car to head to the television studio. My brother and my other relatives insisted on following us in another car for protection. As we began our drive, I noticed that the streets seemed almost empty. There were no traces of the usual Cairo traffic, no traces of security forces. All I saw were army tanks stationed at critical points on main roads.

Our car was stopped twice on the way to the studio. The first time we were stopped by a popular committee in Mohandeseen, where it was clear that the inspectors were young local residents. They asked for our licenses and IDs and quickly let us pass. The second was at an army checkpoint, where only the driver was asked for his license and ID. To me, it all resembled a Hollywood movie. I had not yet realized the magnitude of the change that was taking place in Egypt.

"Were you tortured at State Security?" Mostafa asked me.

"The torture was psychological, but of course I was kicked a few times—by guards, not by officers," I answered.

"Do you plan to mention this?"

"I promised the officers not to get into details and to say that I was generally treated well, which is relatively true, considering the physical horrors that others have experienced when captured by State Security. This is not what concerns me anyway. My concern is to show people how the emergency law facilitates arrests without accusations and how young men are snatched from the street and disappear while their parents are told nothing. The important thing now is for you to tell me what happened while I was gone. I want to understand so I don't say anything wrong!"

The guys said that Tahrir Square was now experiencing its most trying days since the beginning of the revolution. The president had won the sympathy of millions of Egyptians with his second speech. Tens of thousands took to the streets to demand that the protesters evacuate the square. The same segment of the population that would never have dreamed that public opinion could force a minister out of office had suddenly seen the president appealing to them on TV. One sentence from the president's speech in particular had struck home with many. Mubarak said he was born in Egypt and would die in Egypt, the Egypt that he had fought for as a fighter pilot. He also said that he would implement all of the people's legitimate demands. Outside Tahrir Square, in hundreds of thousands of homes, he had touched a lot of hearts.

Yet the square remained to a large extent unaffected by that emotional speech, and those at the sit-in insisted that Mubarak's removal was nonnegotiable. The regime had committed unforgivable crimes over the many years of its rule, and particularly during the past few days. Hundreds of Egyptians had been killed defending their dreams. The situation was irreversible now. Mostafa al-Nagar spoke of his meeting with the recently appointed vice president, Omar Soliman, who refused to consider that the president might step down. His opinion was that the only way forward was to take the revolution to the end. He added that Omar Soliman had played a major part in my release from detainment.

My lack of knowledge of events, which I had perceived as my main point of weakness, turned into my greatest strength during the interview. Most protesters who appeared on television during those days were angry, which is understandable, since many of them had witnessed young Egyptians dying from bullet wounds or being arrested and tortured. But most of the Egyptian public had not lived through the revolution with the protesters, and media outlets were still largely controlled by security forces. The image of the revolution presented by the media at that point was very negative. Rumors and lies about the Tahrir protesters were constantly being disseminated. A vicious Internet attack had been launched against many Jan25 activists, myself included. Some comments on "*Kullena Khaled Said*" and elsewhere labeled me an agent of the United States and Israel.

This was all new to me. It made me dazed and disoriented more than angry.

On the air, I told Mona about my love for Egypt and about the pain of my experience. I was enraged by what had happened to the young protesters, and I was furious that I faced accusations of treachery. I was appearing on air to remind people of the reasons we had taken to the streets on Jan25. My absence from the scene had left my emotions of that day undimmed. I still felt that same raw spirit. I said that Egyptians had taken to the streets to demand the bare minimum of rights we ought to enjoy. We lived in a country where we had lost our sense of belonging. I mentioned the hundreds of thousands of Egyptians who entered the American green card lottery to realize their dream of leaving Egypt.

I mentioned my father, who had cried over my disappearance although he had never been one to cry, and how he had assumed that I was dead and would never return home. I said that even though I had not been physically tortured during detainment, my abduction and arrest without a trial was itself a terrible crime. We still had to fight such crimes, I asserted.

I stressed that I did not want to be treated like a hero. I was only one member of the revolutionary masses who had fulfilled his duty toward his country. It was easy to write, rant, and mobilize people using the Internet. The real heroes of this revolution were the people who had died and been injured. My sacrifices could not be compared to theirs. I was sleeping at State Security while others were engaged in a historic battle.

I was exhausted and had not slept for a long time. The interview was long, and Mona did not interrupt me. At times I lost my ability to think and speak and sat quietly for a few moments to take a few breaths. Mona was very helpful on that front. We spoke calmly, which many viewers found reassuring. I was collected and reasonable, but I remained highly emotional.

Mona asked my opinion about the events and what should happen next, and I stressed that I could not express an opinion when I still knew very little. But I implored everyone not to think of personal gains at this point. This was not a time for promoting, let alone imposing, any ideology or for settling old grievances. Egypt had to come first, and nothing must stand in the way of realizing our dreams.

Mona wanted to mention that some businessmen (who chose to remain anonymous) had decided to donate money to the families of the revolution's martyrs. This is when the director of the show decided to display a video of the martyrs' pictures. I found myself looking at one photo after another of the young Egyptians who had given their lives for a better Egypt. I began to hurt. I hung my head and tried to hold back my tears, but my hoarse voice gave me away. Mona was still talking, but I was no longer listening. I was overwhelmed by grief for the death of those heroes, and by rage at the inhuman murderers who had taken their lives. I raised my head once again and saw the image of the martyr Ahmed Ehab. I did not know him, but his features seemed familiar. He was just another Egyptian who had dreamed of change and paid for it with his life, only two months after his wedding.

Finally I broke down. Mona asked me to stop crying, but I could not. In the midst of my weeping I was only able to say, "I would like to tell every father and mother who lost a child, I am sorry. I am sorry, but this is not our fault, it's the fault of everyone who clung on to power and would not let it go." I stopped, unable to think or speak. This was not how I had wanted to reach people. So I said, "I want to leave," and abruptly ran out of the studio.

Mona left her seat and came running after me. Outside, I found Amr Salama, who immediately reached out to hug me. Everyone around me was crying, including Mona, who said in a husky voice, "It's not you who should cry. It is they who should cry. It's not fair what they did to the youth of Egypt. Please stop." We went to the studio lounge, and I started frantically kicking the ground and the furniture, shouting, "Animals, animals, animals!" What I had gone through in detainment was nothing compared to what I felt at that moment, after seeing those pictures. I wanted justice for those who had given their lives.

My cell phone started ringing as soon as I left the studio, and it did not stop. Even my friends outside Egypt called me to express their solidarity. They had seen the show and they were hurting too. Once I got home, I went online and found that Twitter and Facebook were buzzing with discussion about the show. I had received 1,200 messages at my personal Facebook account, and I tried to read as many of them as I could. The messages were very moving. They were full of love and appreciation. Many of them asked me not to feel guilty. The ones that particularly caught my attention came from people who said they had opposed the revolution up until they saw the show and that now they had decided to go out to Tahrir in support of the protesters.

When I called Ilka, she sounded worried as well. I apologized and asked about the children. She said they were both fine and sound asleep. She asked when I would return to Dubai. At the time, it seemed like a strange question. I told her it would be impossible for me to return anytime soon, considering what was happening. The revolution's goals had not been reached yet. She still thought that I should have learned my lesson after the crisis I had put myself and my family through, while I still wanted to do what I believed was right.

Her anger and fears were justified, of course. I spoke to her for over an hour, trying to calm her down. If I left the country now, I said, I would be betraying the many martyrs who had died for their freedom. I simply could not leave Egypt until the revolution's goals were achieved. I told her about those moments during lockup when I tried and failed to keep myself from thinking of her and the kids so I would not become depressed. I told her how guilty I had felt. Finally, I asked her to let me speak to Isra and Adam in the morning, because I missed them very much.

After Ilka and I hung up, I received a call from an unidentified private number. It was the minister of interior, Mahmoud Wagdy. He said he had watched the whole episode of Mona's show with his family and they were all strongly moved. He expressed his respect for me and for my patriotism and dedication to the cause. I must admit,

I began to trust him a little more after this phone call, which he was under no obligation to make.

It was clear that State Security was counting on me to pressure the protesters to abandon the demand that Mubarak step down. They hoped that the officials' decent behavior and the brainwashing would do the trick.

There is no way I will let anyone pressure me . . . My personal position will only be announced after I've spoken to many people in Tahrir . . . I've been blindfolded from reality for 12 days.

11,093 Likes 6,335 Comments 1,307,026 Views

My interview with Mona al-Shazly spread quickly, thanks to the Internet and social networking sites. Many international media sources, including CNN and the *Guardian*, translated it as well. Numerous journalists tried to portray me as the revolution's champion—my story fit the image. Someone even created a Facebook page called "I Nominate Wael Ghonim to Speak on Behalf of Egypt's Protesters." The page drew about 250,000 members in forty-eight hours but I did not like being promoted as an icon. I thought it was likely to do more harm than good. While the media may have found in my emotional interview just the right dramatic scene for a big story, I continued to remind myself that I was not a hero. The heroes were those brave young men and women who had risked their lives for their country and ended up either martyred or severely injured by Mubarak's brutal regime.

After no more than three hours of sleep, I was up early and reading the comments on the Facebook page. A rumor had spread that the murderers of Khaled Said had escaped, and many members were urging whoever might find them to exact revenge there and then, on the street. I found the idea disturbing. We were fighting for the rule of law and respect for human rights. Any suspect has the right to defend himself in a court of law. So I posted a message:

> There's a rumor that the suspects in the Khaled Said case have escaped from prison. It's bothering me that many people are calling for their immediate execution without a trial. Please. None of us should appoint ourselves judges and issue verdicts against anyone. This is not our role to play. Let us not let excitement and anger cause us to lose everything positive we have achieved so far. Believe me, Egypt comes first and rises above all. Please. Let us all keep that motto in mind and let us keep a clear conscience at all times.
>
> 6,149 Likes 1,763 Comments 1,193,389 Views

I called Amr, Mostafa, and Mohamed Diab to arrange to go to Tahrir together. News started coming in that the square was particularly packed that Tuesday and that my emotional TV interview was one of the reasons behind that. My brother and some relatives came along as well. They suggested that I try to hide my face with a cap to avoid a lot of pushing and shoving if the protesters identified me. I was exhausted, and the burden of my now critical decisions was draining. I felt a huge responsibility with every little choice I needed to make. There was a very heavy and unwanted load on my shoulders.

We arrived at Tahrir Square, which had become much more organized since I had seen it last, on the night of Jan25. There were inspection points at the entrances. The young men and women asked for IDs and apologized for delays as they searched everyone. The square was packed. We could barely see where we stepped. All three of my friends were extremely pleased and said that it had never been that crowded before. The protesters were all calling for the president to step down.

Some people were able to identify me, but my friends convinced them to let me reach the main stage so I could address everyone. That stage was located in front of a number of shops. One shop owner had volunteered his space as a control room where speakers could gather and press releases could be issued. It took us about twenty minutes to reach that room through the crowd. Inside, I removed my cap, and many people recognized me and congratulated me on my release.

To my surprise, Khaled Said's mother entered the room. I rushed to greet her. She held me tightly and cried, saying, "My son did not die. You are my son Khaled." I tried to hold back my tears. We had never met before, but I had followed her news closely and heard her voice in the many videos that I had seen and published on our Facebook page. Chills ran through my body. I knew how this afflicted mother felt. She compassionately asked me if I had been tortured or mistreated, almost as though I were her lost son. I told her that all I cared about was getting justice for Khaled, who had become an icon who inspired all of us to revolt. I also told her that Egypt would become free and that Khaled name's would never be forgotten.

When I reached the stage, what stood before my eyes was like nothing I had ever imagined. Some people started chanting, "Wael, Wael," but I asked them not to. Some of them shouted, "Wael, we hope you did not walk out on us. We hope they did not trick you!" Everyone was waiting for me to reveal my position on the issue of the president's removal, which they all believed was nonnegotiable.

I began by saluting the martyrs. I said that they were the heroes, not me. We all needed to seek justice for them. "This is my country. This is our country," I said. "Every one of us who claims ownership of Egypt must be willing to fight for Egyptians' rights, if not for our generation, then for the next." Then I said that Tahrir Square's demands were my demands—that by this point the president's resignation was a demand everyone should endorse. The crowd started chanting against Mubarak: "Leave, leave, leave!"

At the end, I asked everyone to set aside personal wishes and ideologies. It was time for Egypt to come first and to rule over everything else. I closed by addressing Mubarak: "You've done enough, President. You've had your chance."

Never in my life did I sense as much love in one place as I did in

Tahrir Square on that day. It made me feel like I had been freed after being locked up for eleven years, not just eleven days. The sense of solidarity transcended differences of age, class, culture, education, and religion. Tahrir Square had been transformed into a utopian plot of Earth where all people could believe in their dreams. I saw first-hand how Egyptians had risen as one to seek their rights and achieve their aspirations.

I could not stay for long after stepping offstage. The sheer number of people crowding around me was overwhelming. Everyone was greeting me, eager to kiss and hug me. But I had been deprived of sleep for two days. My friends led me out of the square with great difficulty, and we all took a taxi to my family's home.

The unique and exhilarating experience of Tahrir Square revealed the dignity and beauty inherent in the Egyptian people, and clearly contrasted them with the cruelty and corruption of the regime. As soon as I was home, I published a description of my Tahrir Square sentiments.

> There really is a world of a difference between sensing people's love online and actually experiencing it in the real world. God bless you all. The most important thing now is not to betray the martyrs who gave their blood for their dream.
>
> 17,582 Likes 11,023 Comments 1,090,207 Views

Shortly after my post, my brother, Hazem, received a phone call from the office of Anas El-Feky, the minister of information at the time. The minister's assistant informed Hazem that El-Feky wanted me to appear on state TV. I immediately rejected the invitation and told Hazem to pass on the message that I would never appear on a channel that deliberately misled the Egyptian public and spread lies and rumors about the protesters, facilitating their brutal murder.

The assistant called back a few minutes later and said that the minister of information wanted to speak to me directly. I took the phone, and the minister quickly started telling me that state media were radically changing and that their channels were now open for everyone to speak as freely as they liked. I angrily interrupted. "Are you an Egyptian?" I asked. "Seriously, do you consider yourself someone who loves this country and its people? You are nothing but a propaganda minister who spreads lies, and I certainly do not want to appear on a TV channel run by you! People like you should be in prison and not in government positions! You are an integral part of a regime that has killed hundreds of Egyptians in cold blood." The minister didn't try to argue very much. He told me that he understood what I was going through and that it was important for me to act very wisely at this critical time and not cause more damage to the country. I reiterated that I would not appear on his channel, and we quickly ended the conversation.

My headache was now becoming unbearable. I lay down and effortlessly fell into a deep sleep. As happy as I was with all the love and appreciation I had seen, I was not happy that my anonymity had been lost. I had hoped that no one would ever find out about my political activism or that I was the "*Kullena Khaled Said*" admin. People's attachment to ideas is much stronger than their attachment to individuals, who can be doubted and defamed. I had always tried to steer away from the idea of creating emblems and heroes.

Two or three hours later I woke up, still worn out. I logged on to the page to follow up on its activity and to debate with members who were inclined to end the protests and accept Mubarak's promise that he would transfer power after six months. By now the page's member base had reached 640,000.

> We shall not be fooled by a few reforms . . . The blood of Egyptians is not cheap, dear government . . . When the president's grandson died, all Egyptians cried for his loss . . . But the president did not shed a single tear when more than three hundred of his Egyptian children were killed at the hands of his own police force . . . Our brethren's blood is not cheap, Mr. President . . . By God, I swear that their lives will not go to

waste . . . Mr. President, if you think you can now fool Egyptians with a 15 percent salary raise, then you do not yet know us Egyptian youth.

10,037 Likes 3,553 Comments 726,663 Views

At 5 A.M. on Wednesday, February 9, a friend sent me the video of ABC's Christiane Amanpour interview with General Omar Soliman, the vice president. I had not seen the interview, which had been filmed on February 6, and I found it extremely provocative. The VP spoke of how important it was to be calm now and to grant our trust to the current government in order for the peaceful transfer of power to be fulfilled, as President Mubarak had promised. Soliman also declared that the Muslim Brotherhood and some Islamists were behind the revolution and that they were exploiting the youth for their own purposes. When Amanpour asked him whether he had considered the possibility that it was in fact young Egyptians who wanted their rights and their freedom, he insisted he believed that the protests were not solely the idea of young Egyptians and that they had been pushed to protest by others abroad. He added that the sustained protests were supported by foreign states. He rejected the calls for the president to cede power—according to him, the government feared that without Mubarak, the country would descend into chaos. Everyone believes in democracy, he said, but the question is when to implement democracy and whether the people are ready for it. According to Soliman, Egyptians did not yet have the "culture of democracy."

Soliman's statements were an accurate expression of the regime's attitude toward the Egyptian people. To them, Egyptians were not capable of governing themselves. The government derided people's lack of political consciousness but forgot that that was specifically the result of its own policies, which had led to deterioration on every front: economically, politically, and socially. It was a pathetic excuse. Maintaining the current regime meant maintaining precisely this lack of consciousness.

In his closing comments, when Amanpour asked him what he wanted from the opposition, Soliman addressed the Tahrir protesters directly in his broken English: "I want from the opposition to understand that in this limited time we can do what President Mubarak have pledged and we cannot do more, and when new president will come you will have more time to make any changes you want. We can say only go home, we cannot do more than that. We cannot push them by force. Everybody have to go home, we want to have normal life we don't want anybody in the streets. Go to work, bring once again back the tourists, go to the normal life, save the economy of the country."

I was very angry after watching this interview. My familiarity with the Western press helped me see the hidden meanings behind Soliman's statements. He was sending a clear message to those Western governments that had decided, for the first time in decades, to take the side of the Egyptian people rather than of the regime. He was telling them that Egyptians were not ready for democracy and that authoritarian rule was more favorable to Western interests. He was hinting that democracy, at least for the moment, would pose a threat to the West.

It was at this moment that I changed my mind about appearing in the Western media. It had become necessary. I had already received several interview requests from American media sources by e-mail, and I decided to respond to the most recent one, which was from a CNN correspondent. We spoke on the phone soon after, and an interview was scheduled.

The fact that our revolution did not produce an undisputed leader was viewed by some as a drawback and by others as the secret to its success. The pulse from Tahrir Square was the main driver of consciousness and decision-making. There was no use trying to stand against Tahrir's pulse. All of the regime's desperate attempts failed because the officials could not understand or accept this fact. They tried meeting with the established political factions to influence their decisions, but it didn't work. The protesters stayed in the square and the sit-in continued. The regime was facing an angry generation, a generation that was not politically savvy. It was a generation of people who were not willing to accept middle-of-the-road solutions and who were prepared to sacrifice their lives for the sake of their goals.

> Victory will be ours because we have no hidden agendas . . . Because we do not understand politics, compromises, negotiations, and cheap tricks . . . Victory will be ours because our tears are heartfelt . . . Because our dreams are legitimate . . . Because we have a natural capacity for love . . . Because hope has now established a stronghold inside of us . . . Victory will be ours because we prefer to die an honorable death than to live without dignity . . . Victory will be ours because Egypt comes first and rises above all.
>
> 9,064 Likes 3,571 Comments 1,062,408 Views

Throughout these days, the martyrs' images never left my mind. I posted a suggestion to hold an honorary funeral procession for our martyrs. The procession would take place on Friday, February 11, starting at Al-Nour Mosque and the cathedral in Abbasiyya and marching to Tahrir Square. The idea was accepted by the page members, but it did not spread as widely as Jan25 had. Current events had become extremely heated, and I realized that the ability to mobilize people through the Facebook page would not reach its previous peak of January 28. The momentum had been transferred from the virtual world to the real world, and the virtual world would from now on serve, at best, as a commentator, with limited influence.

I called Ilka and openly talked with her about the risks that I might encounter, and I told her that I had decided to grant her power of attorney over all my assets, including my bank accounts. The upcoming struggles would be unpredictable, and now that I was in the spotlight, I faced greater risks than at any previous time. She asked me to stay safe for her and the kids, and I promised that I would do my best.

Later that day, the legal document was issued, and I asked a friend of mine to ship it to my wife in the UAE.

At noon on February 9, the CNN crew arrived at my family's house and we started filming an interview. The interviewer directed the conversation toward the preparations for Jan25, but I insisted on commenting on Omar Soliman's recent statements. I stressed the fact that the vice president's remarks were baseless. The Muslim Brotherhood and other "Islamist" groups had played no role in preparing for Jan25. The bottom line was that Jan25 was not the work of any political groups. It was a reaction from a generation that had been raised amid fear, failure, and passivity, a reaction mainly inspired by the events in Tunisia.

The interviewer asked my opinion about the current situation. I said that we had been ready to negotiate on Jan25 but that the regime's "negotiation" had come in the form of rubber bullets, water cannons, and tear gas. At this point there could be no negotiation until Mubarak stepped down, I declared. I addressed a message to Omar Soliman, saying that we were no longer afraid. The regime did not and would not understand the younger generation. I pulled out the power of attorney I had issued for my wife and held it up to the camera as a way of letting him know I was not joking. We were ready to die for our demands. We were ready to die for our dream of change in Egypt.

I received two phone calls later that day from officials who wanted to discuss possible ways out of the current crisis. One call came from someone who wanted to arrange a meeting with Dr. Hosam Badrawy, and the other came from Rushdy, the State Security officer who had interrogated me during my detainment. The latter said that the minister of interior wanted to meet me that night about an urgent matter. I agreed to attend both meetings, after clearly stating that the demands of the people in Tahrir were my demands—and they were nonnegotiable.

I called a number of friends and asked them to attend these meetings with me. I wanted to have several witnesses to the proceedings, and I also wanted to expand the scope of discussions and opinions. I had heard that hundreds of people had been arrested during the revolution, so I asked my friends to help me compile a list of names of arrested and missing young people to present to the minister in order to help release them from detainment. I felt more compassion toward detainees than ever before, because I knew firsthand the extent of the struggle that detainees faced.

The journalist Khaled el-Baramawy, who ran Masrawy.com operations at the time, and Amr Salama came with me for the meeting with the minister of interior. Inside the ministry we were greeted by one of the State Security officers who had interrogated me during my detainment. Ironically, he was the same guy who had said that we would never meet again! We waited until the minister concluded a prior commitment, then we entered and were introduced to General Hassan AbdulRahman, head of State Security, who had been the unidentified person at my meeting with the minister, Mahmoud Wagdy, on the night of my release.

The minister spoke of Egypt's situation and how impossible it was for President Mubarak to step down. It was clear that he wanted to convince us to abandon that demand. I tried to be honest and direct in my response. The revolution had no leader, I said. It was more like a raging wild horse that would buck anyone who tried to mount it against its will. All three of us strongly rejected the minister's attempts to persuade us. We spoke of the martyrs who had lost their lives in Tahrir Square.

To my surprise, the minister genuinely believed that the ongoing events in Egypt were the result of a conspiracy against our nation. It seemed that the old guard couldn't see the legitimate reasons for the uprising. They were so disconnected that they believed the masses of young Egyptians were being manipulated. They were living in self-denial, trying to feel better about the atrocities they had committed in the country. The minister said that the Muslim Brotherhood had been breaking into police stations and prisons to free prisoners and cause chaos, with assistance from Hezbollah and Hamas. He even said that the Hezbollah flag had been raised on rooftops overlooking Tahrir Square and that it was Hezbollah's men who had fired gunshots and thrown fireballs at the protesters, in order to pin this aggression on the Ministry of Interior. We didn't believe a word of it. If we supposed, for argument's sake, that Hezbollah members were in fact present, why would they wave their flag? What possible benefit could they gain from doing so, especially if word had spread that Hezbollah was killing protesters, as the minister claimed?

General AbdulRahman, the head of State Security, warned us repeatedly that Egypt was in danger and that utter chaos was around the corner.

I asked him, "Do you realize what your officers and soldiers do to detainees?" I unleashed all my anger. "Why do you torture people and sometimes even electrocute them to death? Why was I arrested, handcuffed, and blindfolded for eleven days, for no reason other than the fact that I demanded the rights of Egyptians, their dignity, and their freedom?"

He appeared moved by my emotional outburst. He apologized to me for what I had experienced and said that the entire strategy of State Security would change now. They had only done these things to defend the future of Egypt and the Egyptians, he said. In his opinion, I needed to rise above my personal grievances during this critical moment and think of what was best for Egypt. General AbdulRahman told us about his personal life, mentioning his children, his MA, and his work on a PhD dissertation in law. He emphasized that many lies were being spread about State Security. He claimed that he was now very happy to see so many proactive young Egyptians who were ready both to vote and to assume leadership of their own country, and that State Security no longer had to fear that extremists would win the elections because Egypt would no longer have the low turnout it had always experienced in the past.

Our conversation was relaxed at times and tense at others. The minister tried to remind us of Mubarak's military background and his service to the country. There it was again, the tired argument:

Mubarak is an elder leader who should not be humiliated. Amr Salama repeated that protesters in Tahrir would accept nothing short of the president's resignation and that Mubarak would be humiliated much more if he were to remain in power for the coming six months. The details of how he would resign could be discussed, but the resignation itself was nonnegotiable. And even if, hypothetically, those of us at this meeting agreed to give up the demand for resignation, it would be impossible to convince Tahrir Square. At the end of our conversation, the minister expressed his concern for the safety of the protesters, especially with news circulating that some of them were planning to march to the presidential palace. He said that the presidential guard would deal aggressively with anyone who posed a threat to the life of President Mubarak. Such a march could lead to a bloodbath, and we had to try to prevent it from happening at all costs.

The situation on the ground was difficult, and Mubarak's regime believed that time was on its side. The protests and sit-ins could only get weaker as days went by, and the regime hoped that protesters would eventually have to give in to the scenario of transfer of power in September, when Mubarak's term in office was up. In light of the ongoing media attack against the revolution, it was clearly a gamble for us to bet on the revolution's comprehensive success. The uncertainty of the final outcome had led many reasonable people to search for a way out of the crisis. In the end, the minister said he knew that Dr. Hosam Badrawy was going to meet with us the next morning and that he had an idea that might actually resolve the situation. He urged us to be willing to listen and cooperate.

We could not make much sense of the minister's closing statements and his promise of a magical solution in the morning, but I was optimistic. In the absence of a revolutionary leader who could easily be pressured, the will of the people was in control. At 3 A.M. we left the minister's office and I went to Amr Salama's house so we could be ready to meet Dr. Hosam Badrawy at an apartment in Zamalek the next morning. First we updated the Facebook page together.

> I feel that very soon we will turn the page, claim our pen, and begin writing our future with our own hands.
>
> 6,316 Likes 2,077 Comments 1,244,267 Views

I sat at Amr Salama's home overflowing with optimism. He asked me if I had seen any videos of the revolution since my release. I said no. He recommended that I watch a video that documented the revolution day by day since it had broken out on Jan25. The video used a thrilling soundtrack together with the most dramatic scenes from the revolution. It started with Jan25, which was the only day I had witnessed from beginning to end. I got goose bumps as I recalled my own experience of that day. Then came images of the clashes from January 28. Security forces used water cannons against thousands of people on top of a bridge near Tahrir Square and targeted the protesters with tear-gas bombs.

Then the heartbreaking images began to appear: Central Security cars running over protesters while they screamed in shock. A protester in Alexandria marching toward the security forces, holding his jacket open, only to be shot and to collapse to the ground, accompanied by the screaming and cursing of the girl who had recorded the incident from her balcony. The Camel Charge Battle, when the Tahrir protesters were attacked viciously by riders on camels and horses amid volleys of rocks and Molotov cocktails. I saw young men carrying the injured to the makeshift hospital, and I realized how limited my role in the revolution really was. I was not a hero compared to any of these men and women. They had lived through something I had seen only in war scenes in Hollywood movies. I started sobbing. I tried to describe my feelings in a post on the Facebook page.

> While I was asleep at State Security and my body did not suffer a scratch . . . heroes were sacrificing their lives for their dream . . . I seriously feel that if I had not been locked up I would have been too afraid to fight alongside you . . . I swear to God you are all heroes . . . I swear to God you are the greatest people

in the world . . . I swear to God you have honored Egypt . . . Believe me it is done . . . Jan25 was the beginning of the end . . . and the end is very near.

8,060 Likes 2,703 Comments 1,459,815 Views

I couldn't sleep and continued to watch that video over and over again. At 8 A.M, Khaled el-Baramawy arrived to meet us and we headed to our destination. Right before leaving Amr's house, I posted a status message in which I attempted to remind myself and all the page members of why I was taking part in all this. It eventually became one of the most read, liked, and commented-on statuses on the page since its inception.

I vow before God that as soon as young Egyptians fulfill their ambitions and dreams, I will completely withdraw from all political life, proud that Egyptians are once again in control of their own destiny. I said this in June 2010 and I am repeating it now, letter for letter. All I want is to walk on the street proud to be an Egyptian.

17,026 Likes 6,937 Comments 2,031,555 Views

As soon as we arrived, Dr. Badrawy initiated the discussion by updating us on the changes he had begun to implement inside the National Democratic Party. He was trying to play a role in leading Egypt to safe shores, he said. He explained that he and some others within the NDP had never been happy with the way party leaders operated, but internal opposition had never been effective. I held firm: the organization responsible for corrupting the political, social, and economic situation in Egypt could not be granted another chance. I was personally determined to prevent the NDP from playing a major political role in the country again. Once again I advised Badrawy to resign and form a new party with his fellow honest party members if he really sought reform.

Dr. Badrawy said that he had accepted the position of secretary-general on condition that he would be able to communicate directly with President Mubarak, and in fact he had been meeting with him, despite the obstacles created by the president's men. In his conversations with Mubarak, he had floated the idea of Mubarak's resigning and delegating responsibility to the vice president. The president had seemed open to the idea, but others around him argued against such an act because they feared for their own interests.

The idea of an honorable resignation, meaning an effective removal from power while nominally maintaining the office of the presidency, seemed to be a reasonable solution at that point. There was too much uncertainty looming, and we needed to prevent any further bloodshed, Dr. Badrawy said. We all expressed our approval of the idea, and both Amr Salama and I began volunteering suggestions for the president's speech. It was important that the term *resign* be clearly articulated and that the president extend condolences to the martyrs' families, we said. He must also express admiration for and approval of the young Egyptians' uprising in order to counterbalance the media's misrepresentation and to avoid any crackdowns against the protesters in the future.

After about an hour of discussion, Dr. Badrawy said he would head to the presidential palace to propose the idea and relay to the president that this was the young protesters' demand. He asked if we would be willing to meet the president in person to present our position. We said that we would, given that we were not negotiating his stay. Dr. Badrawy left and we waited together. Some time later he called us and said that our appointment with the president had been arranged and that the idea of an honorable resignation had been accepted. We were overwhelmed with happiness, thinking that the Egyptian dream was soon to be realized.

In agreeing to meet with a president who had lost his legitimacy, I was not after fame for myself. We knew that we would come under attack once the protesters learned of our meeting. I was sincerely interested in resolving the situation without any further bloodshed, and we were willing to lose our public credibility in return for such a resolution. It should have seemed too good to be true, but I have to admit that in the midst of many momentous events I was a bit gullible. I was searching for a means to realize a dream and was not interested in speculating about conspiracies and political game-playing. I was not after fame for myself.

We were soon on our way, and we were quite optimistic. I called Mostafa al-Nagar and Ahmed Maher and asked them to meet me at a famous hotel near the palace about "an urgent matter." Mostafa was on board immediately, but Maher had something to say first.

"So you fooled me, Wael. We met in Qatar only a few days before Jan25 and you didn't tell me you were the *'Kullena Khaled Said'* administrator!" he joked. "We'll settle this when I see you."

He said that he was at a meeting to coordinate the upcoming Friday protests, but I asked him to leave the meeting and join us as soon as he could.

Amr Salama, Mostafa al-Nagar, Khaled el-Baramawy, and I rode in the el-Baramawys' car. We chanted patriotic songs on the way, thinking that we were about to receive news that would make millions of Egyptians happy. With heavy rain pouring down on the streets of Cairo, it felt like a happy ending scene from a movie.

En route I received a call from the State Security officer that we had spoken to earlier, informing me of a change of plan. We first had to report to the Ministry of Interior, and from there we would go to the cabinet building to meet Prime Minister Shafik before meeting the president. It was a mysterious shift that left Mostafa and Khaled suspicious. But Amr and I thought that there was no reason to assume the worst and that we should keep open minds and proceed. I called Dr. Badrawy to inquire about the change, and he said that the prime minister wanted to listen to what we had to tell the president. He added that we should be as brief as possible with Prime Minister Shafik so that we could head to the presidential palace without too much delay.

We arrived at the Ministry of Interior, still without Ahmed Maher. Rain was pouring down over Cairo, and he was stuck in traffic on the flooded streets. We were asked to follow a car leaving the ministry's headquarters. We did not know who was in that car or who was driving. It was followed by a car full of personal guards. It soon became clear that we were heading toward the prime minister's office beside Cairo International Airport. The traffic was heavy, and all along we continued to sing in anticipation of realizing the dream. We even pictured the moment when Mubarak would publicly announce his resignation and the will of the youth would gain its victory. Just imagining the scene moved us to the point of tears. We chanted a poem by the Arabic poet Abu al-Qasim al-Shabi: "If, one day, the people desire to live free, then fate must answer their call; their night will then begin to fade, their chains will break and fall."

When we arrived at the airport, we left the car and looked to see who was riding ahead of us. It was the minister of interior, Mahmoud Wagdy; his deputy for state security, Hassan AbdulRahman; and Officer Rushdy, who was apparently responsible for coordinating our involvement. We must have seemed so innocent and optimistic to the minister and his party, with their stern looks. I remember even telling Officer Rushdy that our dream was around the corner. He said, "My dream or yours is not what matters. It is Egypt's best interests that matter."

We were ushered into a conference hall with fifty or sixty seats, each with a microphone attached. We didn't understand why we were meeting the prime minister in this particular room, but we decided to wait and see. Magdy Rady, the spokesperson for the prime minister's office, entered the hall and greeted us. He greeted me particularly warmly and spoke about the young Egyptians' impressive achievement.

Then he turned to the issue of the president's resignation. It would create a constitutional dilemma, he said, because the president was the only official with the right to demand constitutional amendments. I smiled and said that legislators had written custom-made constitutional amendments for Mubarak several times in the past without a problem—why should they raise any objections now? I reminded the spokesperson that the protesters would accept nothing short of the president's resignation.

The debate continued until suddenly we heard a rush of footsteps approaching the hall. We all thought the prime minister was arriving and the meeting was about to begin, but we were wrong. A group of at least eight people, roughly our age, entered the hall carrying Egyptian flags. With them was the TV actress Afaf Shoeb, who had publicly called for the protests to end—and a cameraman. It was a trap. We had been set up to stage an image of "dialogue," organized by the government and bringing together proponents and opponents of the revolution, for the benefit of the regime. It wanted to delude Egyptians into thinking that the revolutionaries were now strongly divided and that some of them were actually satisfied and willing to compromise.

We immediately stood up and started to walk out. I turned to Officer Rushdy and said, "Is this how you treat us? By fooling us? You brought us here to film us? You thought we would like to be filmed and appear on television with Their Highnesses?" The minister of interior and the head of State Security approached, but I continued to rant. "You obviously don't understand the youth. You think everything is about deal-making. You thought we were willing to negotiate the rights of martyrs in return for a photo op with a minister or a prime minister."

The minister of interior interrupted, asking me to calm down and let him speak. This was a big misunderstanding, he said. The prime minister had another meeting planned, and these people had come into the hall to wait for him as we had. We did not believe him. But then the minister said that the prime minister had finished his previous meeting, and in fact his office doors opened and several people came out.

Ahmed Shafik appeared and greeted us as though he had no idea what was going on. The minister of interior told him of the "misunderstanding," and Shafik told us that he truly had been scheduled for two conflicting meetings. "Please come into my office," he courteously said. "Let us calm down so we can talk together."

Communicating with nods and expressions, my friends and I agreed to calm down and stick to our plan. There was no need to give up everything in a moment of anger. Full of good intentions, we decided to speak to Shafik in the lead-up to our meeting with the president.

Ahmed Maher had not yet arrived, and I called to ask where he was. He was almost there, he said. I chose not to tell him of our "misunderstanding," because I knew that of all of us, he was the most skeptical of the meeting arrangements. I told him we were in a preparatory meeting that should lead to a meeting with Mubarak, after which he would step down from his office. Maher was surprised and had his doubts, but he was only minutes away.

The atmosphere was charged, and I was the participant who was least able to control his temper. Salama was the calmest, and he asked me not to speak at first, to prevent the meeting from becoming too tense. The prime minister began by reiterating his widespread media statements. He warned against the dire risks that confronted Egypt, specifically mentioning the economic deterioration that was promised if the strikes, sit-ins, and protests continued. We would face famine and chaos, he said. He was very polite and rational, and he tried to defuse our anger as much as possible.

Minutes into the meeting, Ahmed Maher arrived. Greeting everyone, he made an effort to look nonplussed and kept it up throughout the meeting. He insisted on not saying a word. Mahmoud Wagdy even asked him why he would not speak, but his curt answer was that he had nothing important to say and that he preferred to listen.

Amr Salama explained to the prime minster why it was important for the president to step down. He said that Egypt's future was indeed at risk, but the threat came from the government's lack of responsiveness to the people's demands. His logic was brilliant and his manner was nonconfrontational, but after a while we realized that the discussion was going around in circles. Ahmed Shafik did not once mention our prospective meeting with the president. Mostafa whispered in my ear, "We have a big protest to plan for tomorrow. We must leave here, because there is no point in continuing this discussion."

I interrupted the conversation and politely addressed the prime minister: "When will we meet the president?"

Seemingly surprised, he asked, "Are you supposed to meet the president? I have no information about such a meeting."

I doubted that that was true. At that point it became clear that we had indeed been lured into a media stunt that would benefit the current government.

"That was our agreement with Dr. Hosam Badrawy, and he said you knew about it," I replied, maintaining my calm demeanor.

Ahmed Shafik repeated that he had no knowledge of such a meeting. I pulled out my phone and told him that I would call Dr. Badrawy. I turned on the loudspeaker so everyone could hear, but there was no answer. My temper started rising again as I thought back to Mostafa's and Khaled's speculations in the car. They had said that politics was a dirty game and we must not let appearances deceive us. We should not assume that the other party had good intentions.

I tried calling Dr. Badrawy again, and he answered this time. He assured me that a meeting with the president was indeed arranged but said that he had left the presidential palace and had not been informed of any cancellation. I asked him to speak directly to Ahmed Shafik, which he did. Their conversation was dry and brief, and when it was over, the prime minister reiterated that he was unaware of such a meeting.

Maher and al-Nagar got up to leave the meeting, and the rest of us followed them. Al-Nagar apologized for having to end the conversation but said there were important preparations to be made for the following day's protests. We all left the room. The first person out was Maher, and I rushed after him to apologize. I told him that this meeting had made it obvious that all the ministers cared about was holding on to their positions of power.

In contrast to our excitement when we entered the building, we left in great disappointment. I told Officer Rushdy that we were intent on finishing what we had started and that confrontation was now inevitable. The outcome would be either victory or death.

On the way back, Salama and I discussed how gullible we had been to believe the manipulative farce that Dr. Badrawy had led us into so smoothly. I began to realize that many of the conspiracy theories that

I kept hearing about may actually have been true. Tyrants maintain power by conspiring against the will of their own people. We ranted in disappointment and anger, while Mostafa and Khaled reminded us of what they had said all along: this regime was not going to change overnight and suddenly start caring more about Egypt than it did about its personal interests.

Halfway back, Khaled received a call asking us to turn on the radio quickly. We tuned in and heard an announcer cite Dr. Badrawy as the source of the news that the president had decided to step down. He would deliver his speech in just a few hours.

We were astonished. No one understood what was happening. How could the situation have evolved so rapidly and illogically? We tried to think through the events again. Perhaps the meeting with the president didn't take place because he had already decided to step down? I tried calling Dr. Badrawy, but either his phone was busy or he was not picking up.

Meanwhile, international news sources continued to broadcast reports of the president's forthcoming resignation. Perhaps we had been playing a side game with lower-level officials while the real action was happening among Mubarak and his closest advisers? We couldn't figure it out. But there was only one thing to be done: celebrate. We started shouting, "Mubarak is gone! The Egyptian people are free!" We climbed halfway out of the car as it traveled on the bridge, waving flags and yelling the news to passing cars.

When I got home, I started to follow the news on television and tried to unravel the seeming mystery. I saw President Barack Obama interrupt a speech at a Michigan university to congratulate the Egyptian people on achieving their goal. "We are following today's events in Egypt very closely. We'll have more to say as this plays out. But we are witnessing history unfold. It's a moment of transformation that's taking place because the people of Egypt are calling for change," he said. This was, to me and many others, a clear sign that the resignation was sure to happen. It was just a matter of hours.

Everyone in Tahrir Square was cheering, and news channels

started asking people there to comment on the to-be-confirmed news that Mubarak was stepping down. Al Arabiya asked me for a phone interview to congratulate the Egyptian people. I said we had finally succeeded in realizing the objective of the president's resignation and it was now time to go home and think of rebuilding Egypt.

Meanwhile, the Supreme Council of the Armed Forces (SCAF) issued what it called Statement Number One. The statement supported the legitimate demands of the people and their right to hold peaceful demonstrations and added that the SCAF was holding a continuous session. Again, we were not certain that we understood what the SCAF statement meant, but we realized that a lot was happening behind the scenes. The statement was a message of support for the revolution from the army to the people. There was also an indirect message embedded in seeing the SCAF meeting without its supreme commander, who was still officially President Mubarak. It was a sign that Mubarak was losing power.

After speaking to Al Arabiya, I decided to watch the resignation speech from Tahrir Square. I had spoken to many people there who were very happy with the news and eagerly awaiting the speech.

Mubarak began his speech much later than expected, and a wave of suspense washed over us as we waited. But he spoke in his traditional manner. He reminded us of his role in war and in peace, and of his achievements and history. And although he spent a moment offering condolences for the lives of the martyrs, it wasn't half as emotional as the moment when he expressed his pain at Tahrir Square's depiction of him. He even referred to the martyrs as "your martyrs," unwittingly reinforcing his clear disconnection from the square. Mubarak invited everyone to give precedence to Egypt's interests over everything else. He said he had lived on Egyptian soil and would die on Egyptian soil. Then, during the final thirty seconds of the speech, Mubarak said he delegated his authority to Vice President Omar Soliman. It was not actually a resignation from power.

In Tahrir Square, thousands of disappointed yet determined people started chanting, "Leave means go, in case you did not know!" A group of people in the square gathered around me. "What will we do? This is unbelievable. Don't these people understand?" they exclaimed. I had nothing to say to them except that we were the stronger ones and would not give up until our goal was reached. We desired a real resignation and not a murky delegation of power. Disappointed, I left Tahrir with a friend who drove me home.

Two surprising pieces of news surfaced when I met my friends to discuss our next steps. The first was that the Middle East News Agency reported that I had issued a statement inviting all Egyptians at Tahrir to return to their homes following the president's speech. The second was that hundreds of angry protesters had begun marching from Tahrir Square toward the presidential palace to force the president to step down. It was an extremely dangerous development. Everyone was talking about how this could lead to a bloodbath. The presidential guard would attack anyone who approached the palace. I was devastated by this news and extremely apprehensive about the consequences.

I accessed the Facebook page to post a denial of the statement reported by the Middle East News Agency, which had spread on the news tickers of all state media. It was a deliberate plan to use me to influence the protesters.

> Important clarification: I spoke to various TV channels four hours before Mubarak's speech. I have not yet made any comment on the speech itself. I am stating this now because there are multiple news sources that have falsely claimed that I had told people to return to their homes after Mubarak's speech.
>
> 7,323 Likes 7,385 Comments 1,589,234 Views

My online denial was not enough. I received a call from the TV host Amr al-Laithy, who asked that I call in during his show and declare my opinion of events and of the president's decision. I accepted.

On the air, I clarified that I had not made any statements and that, like everyone else, I was not happy with Mubarak's speech. The president was not resigning but delegating authority to his vice president while remaining president. Amr al-Laithy asked what I thought of the march toward the presidential palace. I replied that my personal position did not matter, because the revolution had no leader and no one could influence the protesters who decided to march. He repeated the question in a different way. "Will you march with them?" he asked. I said no, and added that many of us would head to Tahrir the following day to organize a celebration of the martyrs. He then asked me to say something to the people marching toward the palace. I said that it was not for me to address the protesters and asked him and others to stop trying to influence them, because the protesters were not watching television. My final comment was that I would formulate my position after completing an opinion poll on the Facebook page, asking members how they evaluated the events on the ground, as I had done before.

A huge burden of responsibility lay on my shoulders. It was true that my statements might not influence the masses and that rallying through the Facebook page might not be as effective as it had been, since heroes on the ground were now the ones who controlled the course of events. Yet I still sensed—for the first time—that my computer keyboard had become a machine gun, firing bullets with every keystroke. I felt a new sense of responsibility for everything I wrote, and I became very tense.

I designed a questionnaire to canvass people's reactions to the president's speech and find out what they thought the next steps should be, including the idea of marching toward the palace. My plan was to base my own decision on the majority's opinion. We would then all shoulder the responsibility for our actions, even if the result was gunfire from the presidential guard that claimed the lives of hundreds of young Egyptians.

Moments after posting the questionnaire, the polling service that I had been using for months stopped working. The main server that hosted the service suddenly froze because of a burst of user traffic.

Thousands, or maybe tens of thousands, of website visitors wanted to participate and share their opinions, which created more traffic than their server could handle. I tried several times to revive the server, without success. I announced the failure of the polling service on the page as some members accused me of cowardice, lies, and deception. At that point I was physically worn out beyond imagination. I decided to go home, not knowing what I would post next.

An old friend called me to rage over the phone. He said that my lack of support would make me responsible for the death of any protester at the presidential palace. He thought it was my duty to promote the march to the palace; otherwise, the limited number of protesters would make them an easy target. I was extremely disturbed by his sharp tone. We were living through a moment of immense magnitude, and I was sleep-deprived, psychologically tormented, and overwhelmed by the weight of my responsibilities. It was paralyzing.

I entered my family home with my emotions transparent on my face. My mother asked our relatives, who lived in the same building, to come to my room and try to talk to me, but I asked everyone to leave me alone. Then I vented through the Facebook page.

> Dear God, you know our intentions best and that we are seeking our rights and our freedom . . . Dear God, inspire us to do what is right and increase the strength of each and every one of us.
>
> 15,095 Likes 19,690 Comments 1,405,565 Views

I tried to think as I lay down in bed, but before I knew it I was fast asleep, after a day full of unexpected events. For the first time since my release from detainment, I slept for more than seven hours. When I woke up, I was no longer confused. I was determined—and angry. I was angry that we had been used to broadcast an image of "dialogue" to undermine the revolution. I was angry that my name had been used in a press statement to dissuade the protesters from continuing their sit-in. Once again I expressed my complete support for the revolution and its demands on the Facebook page.

> I'm with all of you, I'm with our rights and our freedom. I'm with taking all the corrupt people to justice. I'm with ending over 30 years of violation of the basic rights of Egyptians and their dignity.
>
> 6,632 Likes 2,540 Comments 1,538,294 Views

Mostafa al-Nagar called and said that we must make a media appearance to tell the story of our deception the night before. He suggested appearing with the famous TV host Hafiz al-Mirazi on Al Arabiya.

Later Mostafa arrived at my house with a formal statement he and other political activists had prepared for the army, containing a number of demands. The statement was based on the army's announcement the day before. It began by saluting the SCAF's involvement and went on to outline our demands: confirmation of the complete resignation of the president and his irrevocable removal from power; dissolution of the National Democratic Party and the free formation of political parties; removal of the restrictions placed on candidate nominations for presidential elections; immediate suspension of the emergency law; new rounds of parliamentary elections in all the constituencies in which the court had ruled the previous ballot results illegitimate; monitoring of all elections by local and international civil society, as well as facilitation of the voting process by allowing all resident citizens to use their national ID card and allowing expatriate Egyptians to vote from abroad; the release of all arrested protesters from detainment; the bringing to justice of all the murderers responsible for crimes against Egyptian youth; the bringing to justice of corrupt politicians and the confiscation of all assets stolen from the Egyptian people; and re-creation of the security apparatus on a base of transparency that would prevent monstrous acts of torturing and terrorizing the citizens.

On our way to Al Arabiya, we learned that the numbers at the presidential palace were still limited but increasing as the time for Friday prayers approached. Meanwhile, Tahrir was flooded with hundreds of thousands of protesters and a march had taken off toward the state television building, which was guarded by army forces to prevent it from being raided. By the time my Al Arabiya interview began, I was angrier than I had ever been. I told Hafiz al-Mirazi, the interviewer, that Egypt was experiencing an immense crisis of trust. I said that the country was like a girl who had been raped continually for thirty years, and as soon as she had obtained a knife to fight back, her rapist was begging her to "dialogue" and give up revenge. I read our statement of demands aloud. Mostafa al-Nagar declared that complying with these demands was necessary because martyrs' blood had been shed.

I told the interviewer how Mostafa had held a dying protester in his arms after the man had been shot in the chest. "Mostafa, are we on the right side?" he had asked. "Will I die a martyr?" Speaking through tears as he held him, Mostafa had replied, "Yes, of course." Then the wounded man let out his last breath.

Then my voice rose. "Our tears are not tears of weakness but a sign of strength. Our tears are stronger than the bullets to our chests from Omar Soliman and his men. I am stronger than Omar Soliman and stronger than Hosni Mubarak." Although I ended my statement in the first person, to me it felt as if the protesters had just uttered those words. There was no way we were going to give up our dream, especially not because of fear.

Outside on the street after the interview, people crowded around us once again. I wanted to depart quickly to follow the updates and access the Facebook page. The scene at Tahrir Square was magnificent that Friday. It was clear evidence that the majority of Jan25 protesters had not been satisfied by Mubarak's mere delegation of powers to his vice president. I soon discovered that national television had broadcast the army's approval of the protesters' demands. They even mentioned specific demands, including the ones in our statement. We became hopeful once again.

> Message to the regime: The people on the streets raise the level of their demands with every passing hour. The current demand that needs to be fulfilled as fast as possible is for the president to step down and leave Egypt.
>
> 5,514 Likes 5,030 Comments 1,013,841 Views

At about 5 P.M., television stations announced that an important statement would be broadcast "shortly." After the experience of the previous night, anything was possible. So we decided not to celebrate just yet.

This time Vice President Omar Soliman appeared on national television, looking very serious. "In the name of God, most gracious, most merciful," Soliman read, "my fellow citizens, in the difficult circumstances our country is experiencing, President Muhammad Hosni Mubarak has decided to give up the office of the president of the republic and instructed the Supreme Council of the Armed Forces to manage the affairs of the country. May God guide our steps."

I could not believe it. The dream had been realized. What was impossible for years had been achieved in eighteen days. I went around hugging everyone in the vicinity: my mother, sister, brother, aunt, and friends. We all started singing the national anthem, raising our voices high. I called my wife and said, "You won't believe it! Mubarak is gone!"

"Are you serious?"

"Yes!" I exclaimed. "He's gone! He's gone! He's gone!" I kept repeating it hysterically.

One minute later I updated the Facebook page:

Congratulations, Egypt! This is the historical moment we have been longing to witness!

15,190 Likes 10,832 Comments 968,496 Views

AbdelRahman's family and friends went to congratulate him during the regular Friday visit to his army base. He had never expected events to move so quickly—no one had. Despite the fact that he hadn't had a chance to be physically engaged in the actual protests, his heart and mind never disengaged for a single moment. On that Friday he felt as emancipated as everyone outside, although he still had about ten months of military service to go.

The rejoicing and celebrations in the streets were incredible. Car horns, fireworks, screams, chants, and applause were heard everywhere. It was a defining moment of Egypt's modern history. The will of the people vanquished the will of the rulers.

I went out on the street surrounded by neighbors and friends. Together we stopped the passing cars and chanted with all our might, "Mubarak is gone! The Egyptian people are free!" Celebrations continued throughout the night. We began at Mostafa Mahmoud Square in Mohandeseen and moved all the way to Tahrir Square, which was now filled with people as never before. There I got on the main stage and recited the opening chapter of the Qur'an, "Al-Fatihah," in tribute to the martyrs, and I called on a young Christian to lead a prayer for the martyrs as well. Everyone started chanting, "Muslims and Christians, we are all Egyptians!"

Those moments in Tahrir Square that night were the happiest of my life. We were ending one part of a long story, a story that will not ultimately end until Egypt once again enters the ranks of the world's leading nations. Yet the first mission was accomplished, and the nightmare of the Mubarak regime was over. We had finally found the Egypt that the regime had tried to convince us was lost for good. It took several thousand people only a few hours to convince millions to join them in the search for dignity and freedom, and that Friday, after only eighteen days, all we could say was "Welcome back, Egypt—we really missed you!"

I returned home after midnight with my head held high. I updated the "*Kullena Khaled Said*" page before I went to bed:

Proud to be Egyptian!

9,413 Likes 2,539 Comments 655,359 Views

Epilogue

As I write these final words, Egypt continues on its journey to a better future. The revolution successfully achieved its first objective, removing the key regime figures from power, thus paving the way for opportunity and hope. Yet revolutions are processes and not events, and the next chapter of this story is only beginning to be written. Inevitably, in the wake of the fervor and unity of the revolution, public opinion has fractured, uncertainties have swirled, and we are still a long way from a fully established democracy. I do not pretend to have a crystal ball that can foretell Egypt's future, but I do believe that Egyptians will never again put up with another pharaoh.

Something changed when Egyptians stood up to the Mubarak regime. That change is not limited to our country or to our revolution—it is happening in many countries in the Middle East and on the streets of many cities around the world. Now that so many people can easily connect with one another, the world is less hospitable to authoritarian regimes. Humanity will always be cursed with power-hungry people, and the rule of law and justice will not automatically flourish in all places at all times. But thanks to modern technology, participatory democracy is becoming a reality. Governments are finding it harder and harder to keep their people isolated from one another, to censor information, and to hide corruption and issue propaganda that goes unchallenged. Slowly but surely, the weapons of mass oppression are becoming extinct.

The Egyptian revolution showed us that the great mass of people who are normally risk-averse, aren't normally activists, can become extraordinarily brave and active when they unite together as one. It was like an offline Wikipedia, with everyone anonymously and selflessly contributing efforts toward a common goal. This is why I utterly refuse to be labeled as a hero or take credit for igniting the revolution. I was no more than a guy with some marketing experience who started a Facebook page that snowballed into something greater than any of its thousands of contributors. For a long time I was unwilling to openly defy the regime while thousands of political and human rights activists were on the streets sacrificing and demanding change. Yes, I was fortunate to be an educated "early adopter" of technology and I am part of the urban middle class, which puts me in a small, privileged slice of the total population of Egypt. But I was quite unengaged when it came to politics—a typically cautious, easily intimidated Egyptian who did not dare protest against the regime. When I created the "*Kullena Khaled Said*" page, the whole point was to connect with others just like me. There were many outspoken, more courageous, and radical activists, and some of the other pages and groups that challenged the regime had larger numbers of followers at first. Yet ultimately it was the great middle of the population that needed to overcome its fears and believe that change was possible. I related to that middle because I was one of them.

Revolutions of the past have usually had charismatic leaders who were politically savvy and sometimes even military geniuses. Such revolutions followed what we can call the Revolution 1.0 model. But the revolution in Egypt was different: it was truly a spontaneous movement led by nothing other than the wisdom of the crowd. One day revolution seemed utterly impossible, and there were just a few people dreaming of change. And then, after the brave people of Tunisia ignited a fire that had been smoldering in the hearts of Egyptians and many other Arab people, the impossible quickly became possible. People who would only post comments in cyberspace became willing to stand in public; then those protesters, among many others, made the great leap to become marchers and chanters, and grew into a critical mass that toppled a brutal and tyrannical regime.

Today empowered young Egyptians know that they are capable of shaping the future of their country, truly believing that it is *theirs*. Nevertheless, we need patience as well as passion more than ever, to help us overcome the many challenges we will face until we successfully rebuild our country. In the coming years, I hope to be able to contribute with others to enhancing Egypt's educational system and providing access to technology to as many fellow Egyptians as possible. Democracy must be irreversibly established so that future Egyptians will only have to defend it, not acquire it. The ultimate feeling of happiness that I aspire to will come only when the ambitious dreams of young Egyptians are fully realized.

The revolution in Egypt would certainly have happened without many of the common faces that are linked to it today—and that is the best point of all. This book is a personal memoir and not a comprehensive historical account, and the few names mentioned in the previous chapters, including mine, certainly have no primacy over others in this leaderless revolution. It has been jokingly said that no snowflake in an avalanche ever feels responsible, and when it comes to Jan25, I couldn't agree more. This was the Revolution 2.0 model: no one was *the* hero because *everyone* was a hero.

We shall never forget all those who sacrificed their lives on our path to freedom, and we will continue to sacrifice so that future generations live in the Egypt that their brave predecessors aspired to. The selfless heroes who died defending the dreams of millions of their fellow citizens will forever be our beacon of light. Indeed, it was their courage, determination, and grace that reminded us and the rest of humanity of a universal truth that many seemed to have forgotten: the power of the people will always be stronger than the people in power.

Acknowledgments

When I was told this book had to be written in four months in order for it to be published on January 25, 2012 (our revolution's first anniversary), I thought it was an impossible feat. However, I was quite fortunate to be surrounded by many amazing people, both professionals and friends, who helped make *Revolution 2.0* a reality.

My friend Mohamed Diab helped with the book's structure, and his creative ideas made storytelling an enjoyable experience for me. Mostafa Hashish applied his translation expertise to convert the first Arabic draft of the book into English. I am also greatly indebted to a good friend of mine and the Egyptian editor of this book—who prefers to remain anonymous—for working closely with me and helping me articulate my story and perspective in English.

I was fortunate to have colleagues and friends who reviewed various drafts of the book and offered insightful remarks. I want to thank Hassani, Perihan, Alfi, Gisel, Ali, Antonia, Yonca, and Gawdat for their constructive feedback. I am also grateful to Bruce Nichols, my American editor and publisher, for adding important final touches that were of great value to the book, and to Liz Duvall, my copyeditor.

To my Google colleagues, among many other friends, who provided continuous support for my wife and children while I was missing: I will forever appreciate and will never forget all that you have done.

AbdelRahman Mansour was a truly selfless patriot. He worked for hours and days on the page and helped me administrate *"Kullena Khaled Said"* throughout our battle for freedom.

I would like to thank my mother and father and my brother and sister for standing by me all my life: no words I can write will ever convey how I feel about all of you.

Finally, I want to thank my truly amazing wife, Ilka, for her selflessness and her unrelenting support throughout our marriage: I'm so proud that you are my partner and the mother of my children.

Index